Other Titles in This Series

Mathematical World • Volume 5

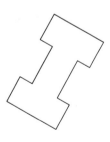

Groups
and Symmetry
A Guide to Discovering
Mathematics

David W. Farmer

American Mathematical Society

1991 *Mathematics Subject Classification*. Primary 20–01, 51–01.

Library of Congress Cataloging-in-Publication Data
Farmer, David W., 1963–
 Groups and symmetry : a guide to discovering mathematics / David W. Farmer.
  p. cm. — (Mathematical world, ISSN 1055-9426; v. 5)
 Includes bibliographical references (p. –).
 ISBN 0-8218-0450-2 (acid-free)
 1. Group theory. 2. Pattern perception. I. Title. II. Series.
QA174.2.F37 1995
511.3′3—dc20

95-21976
CIP

---

10 9 8 7 6 5 4 3 2   01 00 99 98 97

Table of Contents

Preface

This book is a guide to discovering mathematics.

Every mathematics textbook is filled with results and techniques which once were unknown. The results were discovered by mathematicians who experimented, conjectured, discussed their work with others, and then experimented some more. Many promising ideas turned out to be dead–ends, and lots of hard work resulted in little output. Often the first progress was the understanding of some special cases. Continued work led to greater understanding, and sometimes a complex picture began to be seen as simple and familiar. By the time the work reaches a textbook, it bears no resemblance to its early form, and the details of its birth and adolescence have been lost. The precise and methodical exposition of a typical textbook often leads people to mistakenly think that mathematics is a dry, rigid, and unchanging subject.

The most exciting part of mathematics is the process of invention and discovery. The aim of this book is to introduce that process to you. By means of a wide variety of tasks, this book will lead you to discover some real mathematics. There are no formulas to memorize. There are no procedures to follow. By looking at examples, searching for patterns in those examples, and then searching for the reasons behind those patterns, you will develop your own mathematical ideas. The book is only a guide; its job is to start you in the right direction, and to bring you back if you stray too far. The discovery is left to you.

This book is suitable for a one semester course at the beginning undergraduate level. There are no prerequisites. Any college student interested in discovering the beauty of mathematics can enjoy a course taught from this book. An interested high school student will find this book to be a pleasant introduction to some modern areas of mathematics.

I thank Dave Bayer for showing me his method of drawing the Cayley diagrams of wallpattern groups. While preparing this book I was fortunate to have access to excellent notes taken by Hui–Chun Lee and by Elie Levine. It is a pleasure to thank Benji Fisher, Klaus Peters, Sandy Rhoades, Ted Stanford, John Sullivan, and Gretchen Wright for helpful comments on earlier versions of this book.

David W. Farmer
September, 1995

1

Squares, Hexagons, and Triangles

1.1 The square grid. Imagine an infinitely large grid of points.

We can only draw a small part of the picture, so you have to imagine it goes on forever. Think of a giant wall with this as the wallpaper pattern on it.

Now imagine that we took the whole grid and moved everything one unit up. It may help to close your eyes as you imagine this. After shifting the grid up, it looks exactly the same. Wherever there was a dot, there still is a dot, and no dot has appeared where there wasn't one previously. In the diagram above, it appears that moving everything up one space will leave an empty row at the bottom. But remember, the grid keeps going on forever, and there is another row below that one which will move up to take its place, and another row below that, and so on. It is key that the grid goes on forever; there is always another row to take the place of one that just moved up.

Likewise, if we move the entire grid left, right, up, or down a whole number of units, then the grid will end up looking exactly the same. With this in mind we set the following:

Today's Legal Moves:

1) Move the entire grid a whole number of units up or down.

2) Move the entire grid a whole number of units left or right.

3) Any combination of 1) and 2).

1

The important fact for now is: Today's Legal Moves leave the grid looking exactly the same. We say "Today's Legal Moves" because on other days we might choose a different collection of Legal Moves. For example, Today's Legal Moves don't permit us to turn or rotate the grid, but in the future we may permit those moves also.

Now, since the Legal Moves don't do anything to the grid, we have to add something to make it interesting. Suppose we color one square:

Let's see what happens when we apply our Legal Moves. Some example moves are:

a) *Left 2*

b) *Down 1*

c) *Up 1 then Right 3.*

Think of the square as connected to the grid: when we move the grid, the square moves along with it. As we do each of the moves a, b, c, the square moves to the places marked below:

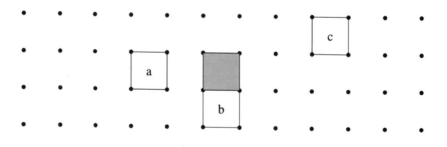

For a), we moved the original square two spaces left. Then for b), we moved *the original square* one space down. That is, we start fresh each time, and all Legal Moves refer to the original beginning position. If you think of the original square as a rubber stamp covered with ink, coloring the square it lands on, then a Legal Move just tells you which square to color. Each Legal Move colors a square, and this leads to the following question:

Question: What happens after we apply all possible Legal Moves to the original square?

There are infinitely many Legal Moves, but we can still imagine they have all been done. If we shade the place where the original square lands each time, then a little bit of experimenting shows that eventually everything gets colored. That is:

Observation 1: As we apply all Legal Moves, the original square covers the whole plane.

That is, just by using Today's Legal Moves, we can move the original square to cover any point on the plane. The result looks like square tiles covering a wall or floor. Continuing to think about tiles, we see that as we move the original square, it never lands on top of a previously placed square. That is:

Observation 2: Different Legal Moves never give overlapping squares.

Actually, that statement needs to be made more precise. For example, Left 2 then Up 1 is a Legal Move, and Up 1 then Left 2 is also a Legal Move, and both of them move the original square to the same place.

Task 1.1.1: Find an iron–clad description of the Legal Moves so that Observation 2 is true.

This problem is very common in mathematics: you notice something, and you have the idea firmly in your mind, but it takes a bit more thought to correctly put your ideas to paper. This process is much like a legislator trying to draft a law without any loopholes. So your first Task is to plug those loopholes! After you do this, try to give a clear explanation of why the squares never overlap.

There isn't anything special about the square we shaded, so we can try shading something else:

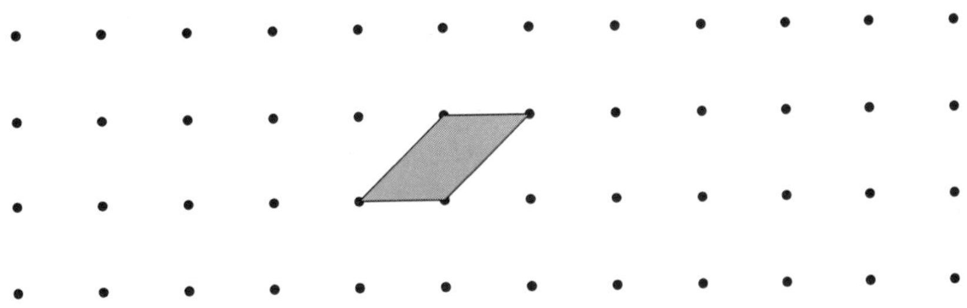

The shape is a parallelogram, but we will refer to it as 'the shape.' Motivated by the facts we discovered about the Legal Moves applied to the original square, we have the following questions:

Question 1: As we apply all Legal Moves, does the shape cover the plane?

Question 2: As we apply Legal Moves, does the shape ever overlap itself?

To answer the questions, we start by looking at the simplest Legal Moves:

a) *Right 1*

b) *Up 1*

c) *Up 1 then Right 1*

d) *Left 1*

Those Legal Moves move the shape to these places:

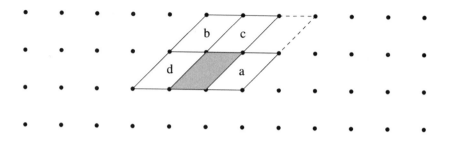

The shape with dashed edges fits nicely, but it doesn't come from one of the moves a, b, c, or d. A bit of experimenting gives that the dashed shape comes from the move Up 1 then Right 2. The different copies of the shape fit together perfectly, leading to these answers:

Answer 1: Yes, the shape covers the whole plane.

Answer 2: No, the shape never overlaps itself.

In some sense, these are the 'good' answers. Answer 1 says that the shape can be used for tiling the floor. Answer 2 says that the shape doesn't have any extra bits which cause the tiles to overlap, or which prevent the different copies of the shape from fitting together properly. Not all shapes have these nice properties. For example, if we start with this triangle:

Applying all Legal Moves yields this pattern:

For this shape, the answers are:

Answer 1: No, the shape does not cover the whole plane.

Answer 2: No, the shape never overlaps itself.

Task 1.1.2: Answer Questions 1 and 2 for various shapes, and find interesting shapes which cover the plane with no overlap when you apply all Legal Moves. Use the examples to devise rules which allow you to quickly answer Question 1 or Question 2. Some sample shapes are given on the next page.

In addition to the example shapes on the next page, you should also invent your own. For example, none of the given shapes has 'curved' edges, so you should try to come up with some that do. The most important goal is to devise rules to help you answer the Questions 1 and 2. Your rules don't have to apply to all shapes: it is useful to have rules which only work some of the time.

Advice. As you go through this book, you may find it helpful to keep a record of your thoughts and ideas. Set aside a notebook for this purpose. Put all of your work there, not just the final answers. It is important to keep a record of the entire process you went through as you worked on a problem, including work which didn't seem to lead to an answer. Your failed method on one problem could turn out to be the correct method for another problem. Having your work in one special place will help you see what you have done and will make it easy to find old work when you need it.

It is important that you spend sufficient time thinking about the Tasks as you encounter them. Some Tasks are easy and some are very difficult, so you should not expect to find a complete answer to every one. If a Task seems mysterious, it can help to discuss it with someone else. Occasionally you may skip a Task and come back to it later, but skipping a Task in the hope of finding the answers in the text will lead you nowhere. The only way for you to find an answer is to discover it yourself. Sometimes this will mean spending a long time on one Task. That is the nature of mathematical discovery. You will find that discovering your own mathematics is not at all like trying to learn mathematics which has already been discovered by someone else.

There is grid paper at the end of the chapter.

1.2 The hexagon grid

We have been imagining an infinite square grid of points. Now imagine the hive of infinitely many bees. The honeycomb would look like this:

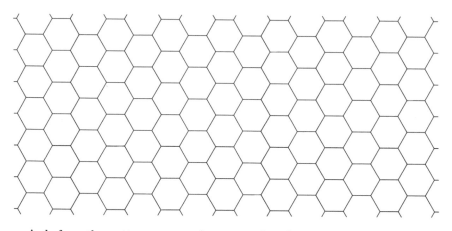

As before, the pattern goes on forever, and we have drawn just a small part of it. We will analyze this pattern in a way similar to our analysis of the square grid. The main steps are:

- Decide on a set of Legal Moves.

- Experiment with what the Legal Moves do to various shapes on the grid.

- Devise rules that determine what the Legal Moves do to various shapes.

We can't use the same Legal Moves as before, because the hexagons are not all spaced a whole number of units apart. However, if we rephrase the previous definition of Legal Move, we get a description that works for both the square and hexagon grid:

Today's Legal Moves:

All ways of shifting the grid up, down, left, or right, or some combination of those directions, so that the grid looks the same before and after the move.

For the square grid, this says exactly the same thing as our previous definition of Legal Moves. The good thing is that the new description also makes sense for the hexagon grid. This process of stating an old idea in a way that makes sense in a new situation is one we will encounter many times.

Task 1.2.1: It was intended that 'diagonal' moves be included in the new definition of Todays Legal Moves. Either explain why diagonal moves are Legal, or else rewrite the new description of Todays Legal Moves so that moves in a diagonal direction are Legal.

Task 1.2.2: Is it true that the new definition of Today's Legal Moves, when applied to the square grid, gives the original definition of Today's Legal Moves? If the answer is Yes, should you change your answer to Task 1.1.1?

Describing a specific Legal Move can be done in a variety of ways. One possibility is to label each hexagon, and then state where each hexagon should be moved. Here is a labeling:

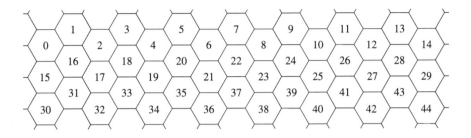

And here are descriptions of two different Legal Moves:

Move A:		Move B:	
	$21 \to 19$		$21 \to 8$
	$22 \to 20$		$37 \to 24$
	$20 \to 18$		$24 \to 11$
	$37 \to 35$		$22 \to 9$
	$10 \to 8$		$33 \to 20$
	$8 \to 6$		$20 \to 7$

You should check that these are sensible descriptions of Legal Moves. There are other ways to describe those moves. For example, Move A could be referred to as 'Left 1 spot,' and Move B could be called 'Northeast 2 spots.' A useful observation is that if we know where a Legal Move sends one hexagon, then we can determine where it sends every other hexagon. For example, in Move B we see that hex 21 goes to hex 8. Since hex 36 is directly below hex 21, it must stay in the same relative position as the grid moves. Thus, hex 36 must move to hex 23, because that hex is directly below where hex 21 goes. Using the same reasoning, we can figure out where any other hex goes.

Now that we have a set of Legal Moves, we can draw a shape on the grid and see what the Legal Moves do to it. The simplest case is shading one hexagon:

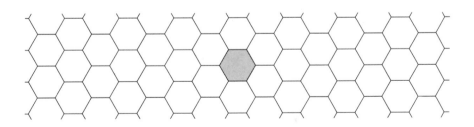

We observed that the Legal Moves are just the ways of moving one particular hexagon to another hexagon, so as we apply all Legal Moves to the shaded hexagon it will cover the whole plane. Also, since the Legal Moves must move a hexagon exactly on top of another hexagon, the shaded hexagon cannot overlap itself as we apply all Legal Moves. In terms of Questions 1 and 2, we have:

Answer 1: Yes, the shape covers the whole plane.

Answer 2: No, the shape never overlaps itself as we apply all Legal Moves.

Here is another shape:

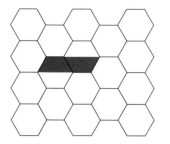 Using various shades for each Legal Move gives this pattern: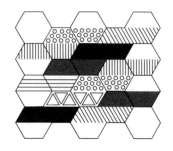

The different copies of the shape fit together in just the right way, so again we have the answers:

Answer 1: Yes, the shape covers the whole plane.

Answer 2: No, the shape never overlaps itself as we apply all Legal Moves.

Our efforts have been concentrated on finding shapes that have the above answers to Questions 1 and 2, and it has grown cumbersome to write out these long answers each time. To make things easier, we give a name to these special shapes.

Provisional Definition. If a shape on a grid covers the whole plane with no overlap as we apply all Legal Moves, then we say the shape is a **basic unit**.

In other words, a shape is a basic unit if the answers to Questions 1 and 2 are Yes and No, respectively. We say *provisional* definition because we may need to state it differently when we move on to new things. The concept will not change, but our description of the concept may need fixing. This process occurred when we needed to change our definition of Legal Move to make it work for the hexagon grid.

Task 1.2.3: For the hexagon grid, experiment with various shapes and come up with some rules about shapes covering or overlapping as we apply all Legal Moves. When possible, use the terminology of basic unit in the statement of your rules.

Task 1.2.4: Determine if your rules for the square grid are also valid for the hexagon grid. When possible, restate your rules so they apply to both grids.

There is hexagon grid paper at the end of the chapter.

1.3 The triangle grid

The third grid we look at is made of triangles:

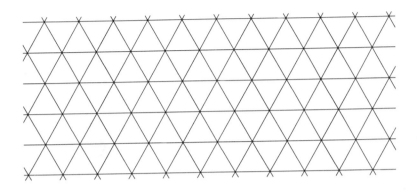

We can reuse our previous definition of Legal Moves:

Today's Legal Moves:

All ways of shifting the grid up, down, left, or right, or some combination of those directions, so that the grid looks the same before and after the move.

Task 1.3.1: Repeat Task 1.2.1.

It is pleasing that our previous definition works in this new situation. To describe specific Legal Moves, we can use methods similar to those used for the hexagon grid. Here is a labeling of the grid:

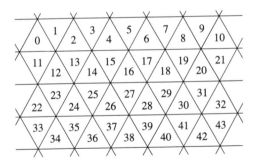

And here are two examples of Legal Moves:

Move A:	Move B:
$37 \rightarrow 25$	$25 \rightarrow 29$
$25 \rightarrow 13$	$24 \rightarrow 28$
$26 \rightarrow 14$	$28 \rightarrow 32$
$39 \rightarrow 27$	$14 \rightarrow 18$
$28 \rightarrow 16$	$11 \rightarrow 15$

Those moves can also be described using terms like 'Northwest 1 spot.' Again we have the observation that once we know where a Legal Move puts one triangle, we can deduce where all of the other triangles go.

Now we can draw a shape on the grid and see what happens when we apply all Legal Moves. For example, if we shade one triangle:

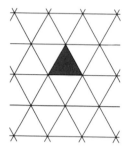

Applying all Legal Moves yields this pattern:

That triangle doesn't cover the whole plane. The shaded triangle 'points up' △, while half of the triangles on the grid 'point down' ▽. If we want to cover the whole plane, then we can try shading a 'pointing down' triangle also.

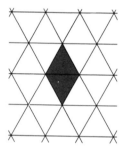

Using various shades for each Legal Move gives this pattern:

We see that the shape made from two triangles is a basic unit.

Task 1.3.2: Is it true that a shape consisting of exactly one 'pointing up' and one 'pointing down' triangle is a basic unit for the triangle grid, even if those triangles are far apart?

Note: The 'shapes' in Task 1.3.2 will be made up of two separate pieces. It is OK for a shape to be made from separate pieces which aren't directly connected.

Task 1.3.3: Find a rectangular shape that is a basic unit for the triangle grid. Do the same for the hexagon grid.

Task 1.3.4: Determine if your rules for the square and hexagon grid are also valid for the triangle grid. When possible, restate your rules so that they apply to all three grids.

There is triangle grid paper at the end of the chapter.

1.4 Putting it all together

Our final goal for this chapter is an Ultimate Rule for determining when a shape is a basic unit for one of the grids. The Ultimate Rule should work for all three grids, it should be foolproof, and it should be applicable to all shapes. We also want an Ultimate Method for producing basic units for each grid.

Task 1.4.1: Given one basic unit for a grid, devise a method of using that basic unit to produce other basic units.

The solution to that Task will provide you with a means of producing a huge number of interesting basic units.

Task 1.4.2: Devise a rule that takes any shape on a grid and determines if it is a basic unit.

Task 1.4.3: Compare your answers to the previous two Tasks. Does there seem to be a connection between them? If not, try to rewrite them so it is clear how they are related.

1.5 Notes

Note 1.5.a: What we call a *basic unit* is also called a **fundamental domain**.

Note 1.5.b: This chapter was about three grids. We stopped at three because there are only three different regular grids in the plane. The key word here is *regular*. Once you have a precise definition of regular grid, it takes only a small amount of geometry to show that these three grids are the only possibilities. For example, if you try to make a grid from pentagons you will fail because it is impossible to fit regular pentagons together in a way that lies flat and covers the plane.

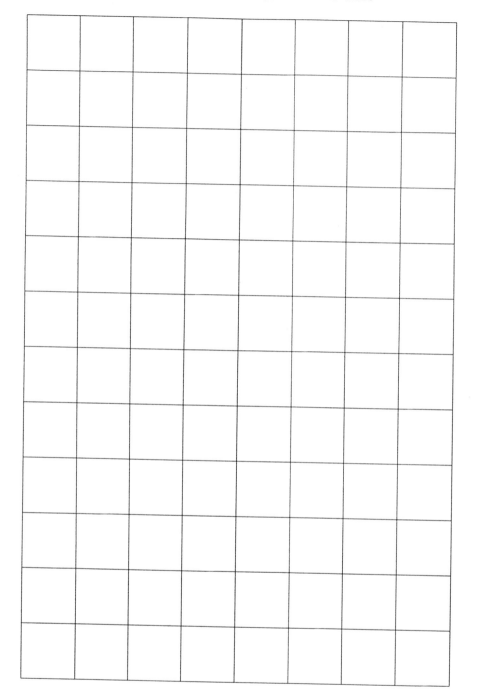

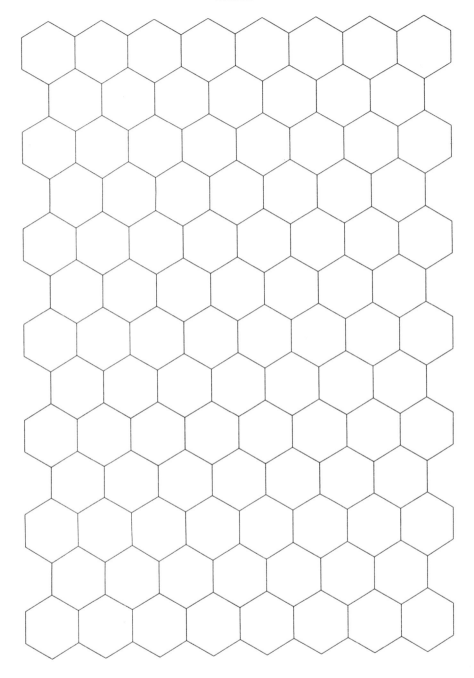

2

The Rigid Motions of the Plane

2.1 Translation and rotation. All of the Legal Moves in the previous chapter involved shifting the entire plane as one rigid unit. There are many other ways of rigidly moving the plane. Finding these other rigid motions, and understanding how they work, is the subject of this chapter.

Preliminary definition. A **rigid motion** of the plane is any way of moving all the points in the plane such that

– The relative distance between points stays the same.

– The relative position of points stays the same.

As before, we say 'Preliminary' because we can't be sure that our first attempt at a precise definition will correctly capture what we have in mind. If this definition leads to unexpected results, then either we accept those results, or else we change the definition.

Instances of the two properties of a rigid motion are: if points A and B are 2 meters apart, then after applying a rigid motion they should still be 2 meters apart; and if point C is halfway between A and B, then after applying a rigid motion it should still be halfway between A and B.

The simplest rigid motion is a *translation*. The Legal Moves in the previous chapter are all examples of this type of motion. In a **translation**, everything is moved by the same amount and in the same direction.

We specify a translation by drawing an arrow. The arrow points in the direction of motion, and the length of the arrow is the distance everything moves.

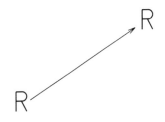

We can also describe a translation by specifying an amount up/down and an amount left/right to move everything. The translation can be by any amount, we are not limited to the whole number translations of the previous chapter.

A translation moves each point
by the same amount, so it does
not change the relative distances
or positions of points, as this example illustrates:

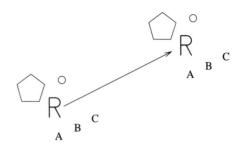

In other words, a translation fits our definition of rigid motion.

Another rigid motion of the plane is **rotation**. A rotation fixes one point,
and everything rotates by the same amount around that point.

To specify a rotation, show which point
is fixed and the amount which everything
turns around that point. The fixed point,
called the **rotocenter**, acts like an axle,
and the two lines we draw to indicate the
rotation act like the spokes of a wheel.

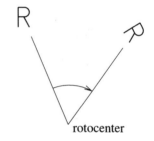

There are several different ways to measure the amount of a rotation. In
this book we will measure rotations as fractions of a full–turn. For example, the
rotation shown above is by one–sixth of a turn clockwise.

A rotation fits our definition of rigid motion, as this example illustrates:

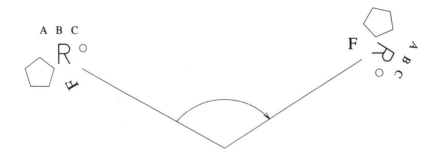

Here is a useful method for drawing a rotation.

First turn the paper so that the starting line of the rotation is vertical, and the starting figure is poised directly on top of the starting line. Fix this picture in your mind.

Next, rotate the page so that the ending line of the rotation is vertical. Draw the figure on top of the ending line so that the result looks like the picture in your mind of the starting position. A bit of practice is needed to get this last bit right.

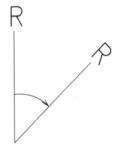

If you look at the examples on the previous page, you will see this process in action.

Another point to keep in mind while drawing rotations is that the two line segments used to indicate a rotation should be of the same length; as mentioned before, those lines are acting like spokes of a wheel. That observation is worth formulating as a rule:

Rule. The distance from the starting figure to the rotocenter is the same as the distance from the ending figure to the rotocenter.

That rule will play a key part in the next section, so you should make sure that you believe it is true.

Task 2.1.1: Practice drawing some rotations. Use a variety of shapes, and do examples of both clockwise rotation and counterclockwise rotation by various amounts. Check that your examples fit the definition of rigid motion and obey the above rule.

2.2 Combining translations and rotations

At present we have two rigid motions of the plane: translation and rotation. Now we look at combining them. The simplest case is when we translate, and then translate again. For example:

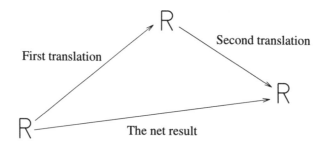

In this example, we see that a translation of a translation is a translation. In fact, this always happens.

Rule. A translation of a translation is a translation.

It is important that you convince yourself the rule always works.

A more complicated situation is a translation followed by a rotation.

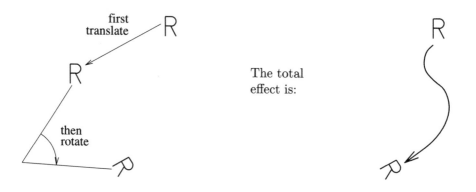

The wavy line indicates that some rigid motion took the first R to the second R. The wavy line does not tell us anything about that rigid motion, it is just a way to show which is the beginning R and which is the ending R. In this example, the rigid motion is a rotation:

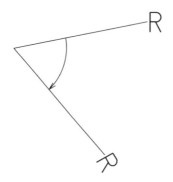

You should spend a few moments checking that the given rotation is correct.

Task 2.2.1: Do some experiments to check that rotation of a translation is always another rotation. Devise a method to determine *which* rotation is the result. That is, find the rotocenter and the amount of rotation for the resulting rigid motion.

Task 2.2.2: Suppose that two R s are drawn somewhere on a sheet of paper. How do you determine which kind of rigid motion, translation or rotation, will move one R to the other R ?

We have not yet looked at translation of a rotation, and rotation of a rotation. This is discussed in Section 2.5

2.3 Mirror reflection

Our list of rigid motions is not complete: another rigid motion is **reflection**.

A reflection is determined by a **mirror line**. We use a dashed line to indicate a mirror.

R ¦ Я

A reflection transforms the English letter R into the Russian letter Я, pronounced 'ya.' If you haven't studied Russian, then at first the name 'ya' may sound funny, but soon you will see that this is much more convenient than 'backwards R.' Say 'ya' out loud a few times until you get used to it.

A reflection fits our definition of rigid motion, as this example illustrates:

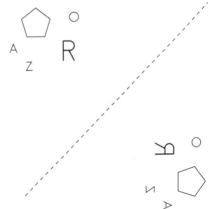

Turning the page so the mirror line is vertical makes it easier to see that the reflection is accurately drawn. The two sides really are 'mirror images' of each other, and if the paper were folded along the mirror line, the original figure and its image would lie on top of each other. This observation can help in drawing the result of a reflection. Two other useful observations:

- Points on the mirror are unchanged by the reflection.

- The distance from a point to the mirror is the same as the distance from the image of that point to the mirror.

These observations will be useful in trying to understand how reflection interacts with the other rigid motions; this is our next topic.

For combining a reflection and a reflection, there are two cases, depending on whether the mirrors intersect or are parallel. An example with parallel mirrors:

We reflect first in mirror m_1, then in mirror m_2. In this example, we see that the net effect of the two reflections is a translation:

Task 2.3.1: Explain why, if the mirrors are parallel, a reflection of a reflection is a translation. Determine how to find which translation occurs. That is, devise a method for finding the direction and the amount of the resulting translation.

If the two mirrors intersect then the situation is slightly different:

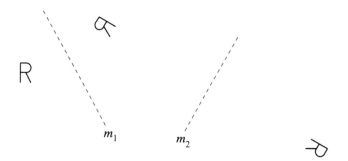

If you extend the mirror lines then you will find the point where the mirrors intersect. In this example, the net result is a rotation:

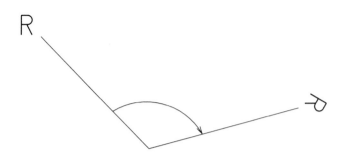

Task 2.3.2: Explain why, if the mirrors intersect, a reflection of a reflection is always a rotation. Determine how to quickly find which rotation occurs. That is, devise a method for finding the rotocenter and the amount of rotation.

2.4 Glide reflection

Next we combine reflection with translation.

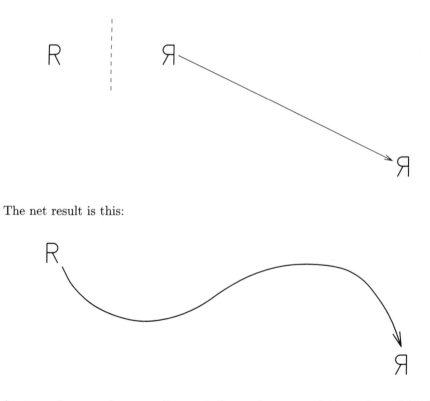

The net result is this:

Again we have used a wavy line to indicate that some rigid motion, which has yet to be determined, moved the R to the Я. A bit of experimenting convinces us that *none* of the rigid motions we currently have is able to move the R to the Я in one motion. It is true that we can combine two rigid motions to accomplish that task, but none of translation, rotation, or reflection suffices by itself. There are two ways to fix this problem. The boring, short–sighted, head–in–the–sand method is to say, "ok, we will have to live with the fact that we need to combine various operations in order to get all rigid motions of the plane." The interesting, far–sighted, take–action method is to *invent another rigid motion of the plane.* This new rigid motion is called a *glide reflection.*

A **glide reflection** is a mirror reflection, followed by a translation *parallel* to the mirror. This counts as *one motion.*

We show a glide reflec-
tion by drawing a mirror,
called the **glide line**, with
a dashed arrow, showing
the translation amount,
immediately next to it:

R   Я

Я

The phantom hazy figure is *not* part of the glide reflection: it is only there to assist in seeing the glide reflection and to help draw it properly. It is usually a good idea to lightly sketch the phantom figure when drawing a glide reflection.

Here is another example:

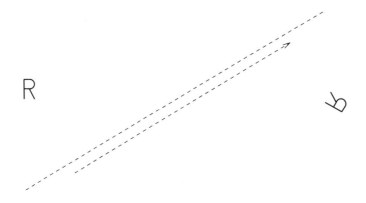

You should sketch the phantom Я to see that the glide was drawn correctly. Hint: the phantom figure will be on top of this sentence.

A glide, like a mirror reflection, transforms an R into a Я. Another similarity with mirror reflection is that the initial R is the same distance from the glide line as the final Я. These observations are important for understanding glide reflection.

Task 2.4.1: Given a translation of a reflection, is there always a glide reflection that accomplishes the same motion? As a start, draw the glide reflection that accomplishes the translation of a reflection that began this section.

2.5 Combining rigid motions

In this section we examine further the various ways in which the rigid motions can be combined. There are quite a few combinations, so it may be advisable to just skim this section on first reading. You can always look back for whatever information you require.

The following chart is a convenient way to classify the rigid motions. Recall that a **fixed point** of a rigid motion is a point which is not moved by the rigid motion.

Translation : Moves R → R, has no fixed points.

Rotation : Moves R → R, has one fixed point.

Mirror Reflection : Moves R → Я, fixes its mirror line.

Glide Reflection : Moves R → Я, has no fixed points.

Do–nothing : Moves R → R, fixes every point.

Those properties can give us some short–cuts when analyzing combinations of symmetries.

Task 2.5.1: Use the properties above to give a quick explanation of why a reflection of a reflection, with intersecting mirrors, must be a rotation.

Now we finish our analysis of translation and rotation.

Task 2.5.2: Does a translation of a rotation work the same as a rotation of a translation?

Task 2.5.3: A rotation of a rotation can sometimes be a translation. Determine exactly when this happens.

Task 2.5.4: A rotation by a half–turn is somewhat easier to analyze than the other rotations:

a) Determine what happens when a half–turn rotation is combined with a translation. That is, find the new rotation amount and the location of the new rotocenter. There are two possibilities, depending on whether you rotate first or translate first.

b) Determine what happens when one half–turn rotation is combined with another half–turn rotation.

Task 2.5.5: Suppose you are given two rotations such that the combination of them is another rotation. Devise a way to find the rotocenter and the rotation amount for the new rotation. A picture like this can help locate the new rotocenter:

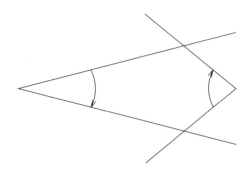

Task 2.5.6: Suppose you draw two R's on a piece of paper, such that one is 'tilted' compared to the other. Then you know that some rotation will move one R to the other. Devise a way of finding *which* rotation does the job. That is, determine the rotocenter and the amount of the rotation.

Task 2.5.7: In the Task 2.3.2 you found that a reflection of a reflection, with intersecting mirrors, is a rotation. Is the reverse true? That is, given a rotation, can you always find two reflections so that first doing one reflection and then the other is the same as the rotation you were given?

We have completely analyzed translations and rotations, so now we move on to mirror reflection and glide reflection.

The first case we have not yet considered is rotation of a reflection. Here is an example:

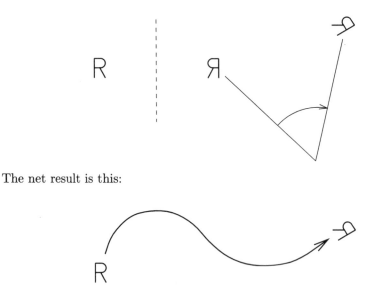

The net result is this:

Again, the wavy line indicates a rigid motion which has yet to be determined.

Since the starting figure is an R and the ending figure is a Я, either a reflection or glide reflection is the desired rigid motion. In this example, it is a glide reflection which does the job:

You should check that the given answer is correct.

The following Tasks will complete our study of the rigid motions of the plane. The first Task shows that a rotation of a reflection is always a reflection or glide reflection. This result is used in some of the other tasks, so if you don't do the tasks in order then you can take this result as given.

Task 2.5.8: Given an R and a Я drawn somewhere on a piece of paper, devise a method of finding the reflection or glide reflection which sends one figure to the other. That is, given the two figures, find the glide line and the translation amount.

Task 2.5.9: What rigid motion do you get when you combine two glide reflections? Note: there are two cases, depending on whether the glide lines intersect or are parallel.

Task 2.5.10: Given a reflection and a rotation, how can you quickly tell if the rotation of the reflection is another reflection, or a glide reflection?

Task 2.5.11: We introduced glide reflection because we wanted each rigid motion of the plane to be accomplished by just one move. Are there any other rigid motions which we haven't discovered yet? Either find another, or explain why there aren't any more.

2.6 Notes

Note 2.6.a: The operation of 'do–nothing' can be considered a rigid motion of the plane. The do–nothing operation just leaves every point in the same place as it started. We will often find it useful to classify do–nothing as both a translation and a rotation.

Note 2.6.b: The phrase *the set of translations and rotations is closed* is a shorthand way of saying, "a translation or rotation of a translation or rotation is a translation or rotation." Similarly, the set of translations is closed. The set of rotations is not closed, because we can combine two rotations to get a non–rotation. More generally, given a way of combining pairs of elements from a set, called a **binary operation**, the set is said to be *closed under the operation* if the result of the operation is again inside the set. In this chapter we are combining rigid motions by first doing one and then the other, so it is understood that this is the operation we have in mind when we say, "the set of translations is closed." In some sense, it is only worthwhile to study closed sets of rigid motions. If you attempted to study only reflections, for example, you would be doomed to failure, for when you combine two reflections you get a rotation or translation. This will become an important part of Chapter 5.

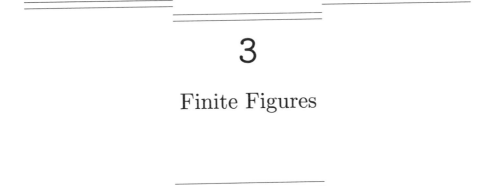

3

Finite Figures

3.1 Symmetry. What is symmetry? These figures fit the usual conception of the word 'symmetric':

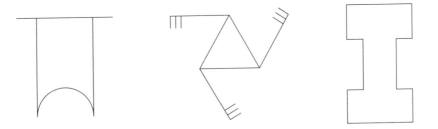

The first figure has mirror symmetry, the second has rotation symmetry, and the third has both mirror symmetry and rotation symmetry. The common idea in each case is that there is a rigid motion which leaves the figure unchanged. We will take this as our definition.

Provisional Definition. A **symmetry** of a figure is a rigid motion which leaves the figure unchanged.

A symmetry is a rigid motion which leaves the figure looking exactly the way it started. The usual meaning of the phrase "the figure is symmetric" can now be rephrased as "the figure has a symmetry." In the previous chapter we found all rigid motions of the plane, so we will concentrate on finding the symmetries of planar figures. Our knowledge of how the different rigid motions of the plane combine with each other will be very useful in this chapter and the rest of the book.

Special Convention. We always count do–nothing as a symmetry of a figure.

The rigid motion which leaves everything in its original spot will always count as a symmetry, so figures which fit the usual conception of 'symmetric' will have more than one symmetry. Sometimes we refer to do–nothing as the **trivial symmetry** of a figure.

Now we list all symmetries of the three figures above.

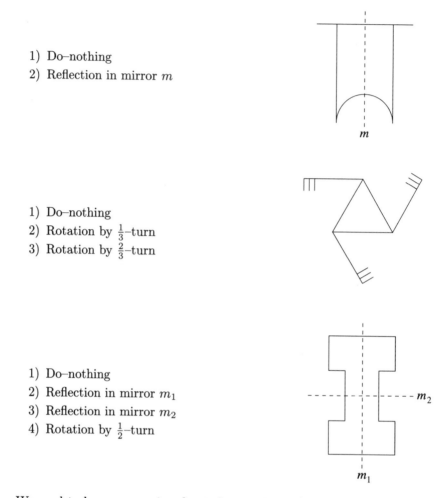

1) Do–nothing
2) Reflection in mirror m

1) Do–nothing
2) Rotation by $\frac{1}{3}$–turn
3) Rotation by $\frac{2}{3}$–turn

1) Do–nothing
2) Reflection in mirror m_1
3) Reflection in mirror m_2
4) Rotation by $\frac{1}{2}$–turn

We need to be more precise about the rotations of the second figure. The $\frac{1}{3}$–turn symmetry is ambiguous, since it could refer to either a clockwise or counterclockwise turn. To avoid ambiguity, we adopt this special convention:

Special Convention. All rotations will be measured counterclockwise.

So the two listed rotation symmetries of the second figure are counterclockwise rotations. There is another possible ambiguity with rotations. Your next task is to fix this problem.

Task 3.1.1: Resolve the following ambiguity: should rotation by one full turn be counted the same as do–nothing?

The examples listed above are *finite figures*, that is, shapes which do not have any translation symmetries.

Definition. A **finite figure** is a shape that has no nontrivial translation symmetries.

The word *nontrivial* is necessary because do–nothing counts as a translation, and all shapes have do–nothing as a symmetry. The grids in Chapter 1 are not finite figures, because they have translation symmetries. In fact, the Legal Moves of that chapter are exactly the translation symmetries of those grids. For the rest of this chapter we will concentrate on finite figures. In later chapters we will study figures with translation symmetry.

Task 3.1.2: Draw some finite figures and find all of their symmetries. Some good examples to try: a triangle, a square, a star.

Task 3.1.3: Can a finite figure have two rotation symmetries with different rotocenters?

Here are the symmetries of a square:

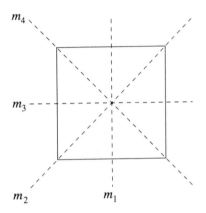

1) Do–nothing
2) Rotation by $\frac{1}{4}$–turn
3) Rotation by $\frac{1}{2}$–turn
4) Rotation by $\frac{3}{4}$–turn
5) Reflection in mirror m_1
6) Reflection in mirror m_2
7) Reflection in mirror m_3
8) Reflection in mirror m_4

If we count do–nothing as a rotation, which makes sense because it is rotation by no amount, then we see that a square has 4 rotation symmetries and 4 reflection symmetries. Is there a pattern? Would a pentagon have 5 rotation symmetries and 5 reflection symmetries? It turns out that one more condition is needed for this pattern to hold true: the pentagon has to be *regular*.

Definition. A polygon is **regular** if all its sides are equal and all its angles are equal.

Here is an example to illustrate the idea:

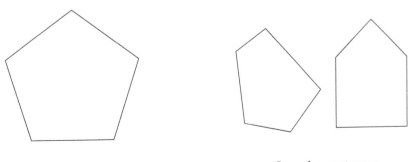

A regular pentagon   Irregular pentagons

Task 3.1.4: What sport makes use of an irregular pentagon?

Task 3.1.5: Explain why the regular n–gon, that is, the regular polygon with n sides, has exactly n rotation symmetries and n reflection symmetries.

Task 3.1.6: Experiment to see what symmetries an irregular polygon can have. Pentagons and hexagons are good examples to try.

3.2 Combining symmetries

Our starting point for this section is the following observation:

Observation. If we combine two symmetries of a figure, then we get yet another symmetry of the figure.

Here we use 'combine' in the sense of applying one symmetry and then the other. This is how we combined rigid motions of the plane in Chapter 2.

The above observation is correct because both symmetries are rigid motions, so the combination of them is a rigid motion, and both symmetries leave the figure unchanged, so the combination leaves the figure unchanged. In other words, the combination is also a symmetry of the figure. Let's do a specific example. Referring to the symmetries of the square listed in the previous section, let m stand for 'mirror reflection of the square in mirror m_4,' and let r stand for 'rotation of the square by $\frac{1}{4}$–turn.' These diagrams explicitly show the effect of m and r :

Task 3.2.1: Make diagrams showing all 8 symmetries of the square.

Let's combine r and m. The notation rm stands for 'rotation of mirror reflection of the square.' That is, first mirror reflect, then rotate.

Very Important Note. The symmetry rm means first mirror reflect, then rotate. Combinations of symmetries are read right to left. This is because rm means rotation **of** mirror reflection **of** the square, so we are rotating the figure which has already been reflected.

Combining the two diagrams above shows the effect of rm :

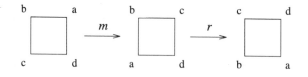

So the total effect of rm is:

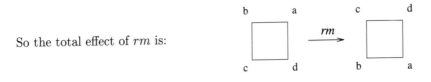

Since rm is a symmetry of the square, it must be one of the 8 symmetries from the previous section. In Task 3.2.1 you made diagrams of each symmetry. Referring to those diagrams you see that rm is the same as reflection in mirror m_3.

The other way to combine r and m is mr, meaning first do r and then do m. Will this be the same as rm? Let's find out. Combining the diagrams for m and r in the other order gives:

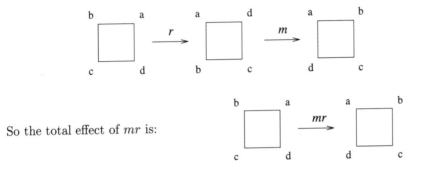

So the total effect of mr is:

Referring to your diagrams from Task 3.2.1, we see that this is the same as reflection in mirror m_1. It is interesting that rm is not the same as mr. The rest of this section is devoted to understanding how the rotation and mirror reflection symmetries of a figure interact with each other. Our plan is to experiment until we see a pattern, and then try to explain the pattern. It is possible, but very tedious, to draw lots of pictures of the various combinations of symmetries. A much better method is to manipulate small models of the figures.

Task 3.2.2: Make cardboard cutouts of a regular triangle, square, pentagon, and hexagon. Label both sides of the figure so that each corner gets the same label on both sides. This means that if the letters *abcde* read clockwise on one side, they will read counterclockwise on the other side.

Here is a way to make a pentagon. Take a flat strip of paper and tie it in an overhand knot.

An overhand knot:

Slowly pull the knot tight and press it flat. Fold over the dangling ends and you will have a pentagon.

Task 3.2.3: Convince yourself that flipping over the cutouts you made in

Task 3.2.2 has the effect of mirror reflection.

Task 3.2.4: Let r stand for a $\frac{1}{2}$–turn of the square, and let m stand for reflection in mirror m_2. Is rm the same as mr ? Manipulating a cardboard square will make this an easy Task.

Hopefully you just did Task 3.2.4, and you found that rm was the same as mr. So sometimes $rm = mr$, and sometimes $rm \neq mr$. To figure out what is really going on, we need to be more organized. First we fix some notation; this notation will be in force for the rest of the book.

Notation.

- 1 stands for do–nothing.

- r stands for the smallest counterclockwise rotation symmetry of the figure. So for the square r stands for $\frac{1}{4}$–turn, for the pentagon r stands for $\frac{1}{5}$–turn, etc. To know what is meant by r, you must be aware which figure is being discussed.

- r^2 stands for doing r twice, r^3 stands for doing r three times, etc. So for the square r^2 stands for a $\frac{1}{2}$–turn, and for a hexagon r^2 stands for a $\frac{1}{3}$–turn.

- r^{-1} stands for the smallest clockwise rotation symmetry of the figure. That is, r^{-1} is the opposite rotation of r. Similarly, r^{-2} is the opposite rotation of r^2, etc.

- m stands for a mirror reflection symmetry of the figure. Usually we need to draw a picture to show which mirror reflection we have in mind.

Task 3.2.5: Write out exactly what is meant by r, r^2, r^3, and r^{-1} for a few figures. Is it always true that $rr = r^2$, and $rr^2 = r^3$, and $rr^{-1} = 1$?

Task 3.2.6: In algebra you have the rules of exponents: $r^a r^b = r^{a+b}$ and $r^0 = 1$. Explain why these rules make sense in our new notation.

In our new notation we can express the four rotation symmetries of the square as 1, r, r^2, and r^3. A useful rule is $r^4 = 1$. For the square, r stands for $\frac{1}{4}$–turn, so this rule merely says that one full turn is the same as do–nothing. Similarly, $r^5 = r$, $r^6 = r^2$, etc. These equations are only true for the square. Similar, but not identical, equations can be found for other figures.

Let m stand for the indicated mirror reflections:

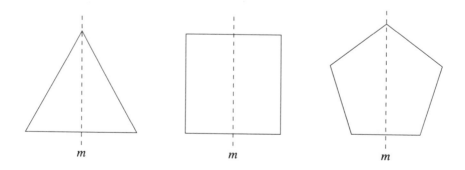

Task 3.2.7: For each figure above, experiment to see that mrm is a rotation. Determine which rotation it is, expressing your answer in terms of r.

Task 3.2.8: The square has 8 symmetries. Five of them are 1, r, r^2, r^3, and m. Express the other three symmetries in terms of r and m. Do the same for the triangle.

Task 3.2.9: The four mirror reflection symmetries of the square can be written as m, mr, mr^2, and mr^3. Draw a square and label the mirror lines associated with each of these reflections.

By now the alert reader will have concluded that the eight symmetries of the square can be represented as $\{1, r, r^2, r^3, m, mr, mr^2, mr^3\}$. Therefore, any symmetry of the square can be expressed as one of those eight choices. For example, r^2mrmr^3mr is a symmetry of the square, so it is the same as one of those eight. The question is, which one? If we manipulate our cardboard cutout, we find the following:

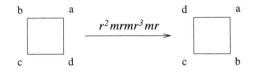

Referring to your answer to Task 3.2.9, we see that this is the same as mr.

Special Convention. We adopt the following standard form for writing the symmetries of a regular polygon: the rotations will be written as 1, r, r^2, etc, and the mirror reflections will be written as m, mr, mr^2, etc.

Given any symmetry of the square, we will rewrite it in our standard form. We need to find a quick and efficient method of doing this; the method we used above of manipulating cardboard cutouts would become tedious if we had to do it repeatedly. Our primary need is for a method of 'moving the m to the left.' For example, rm is not on our list of eight symmetries of the square, but a little fiddling shows that rm is the same as mr^3, and this form of the symmetry has the m where we want it: on the left.

For the next four Tasks it is OK to use your cardboard cutouts. By the end of those Tasks you should have some rules which will let you dispense with the cutouts.

Task 3.2.10: For a few different figures, express rm, r^2m, and r^3m in our standard form.

Task 3.2.11: Verify that $rmr = m$ is true for the triangle, square, and pentagon. Why does this rule work for all three figures, while most other rules need to be modified depending on how many sides the figure has? Is it also true that r^2mr^2 is always the same as m ?

Task 3.2.12: Use Tasks 3.2.10 and 3.2.11 as inspiration to find a rule expressing rm in terms of something with m on the left. Look for one rule that works for any figure, as opposed to different rules for different figures. Then find similar rules for r^2m and r^3m.

Task 3.2.13: Explain why $m^2 = 1$. That is, doing m twice is the same as do–nothing. Another way to write it is $mm = 1$.

The results of Tasks 3.2.12 and 3.2.13, along with the rules $r^4 = 1$ for the square, $r^5 = 1$ for the pentagon, and so on, are all that is needed to reduce any combination of r's and m's to its simplest form. You will find that using those rules is faster and more accurate than manipulating cardboard figures.

Task 3.2.14: Use your rules from Tasks 3.2.12 and 3.2.13 to check the equations below. In other words, simplify the left side of each equation and check that you get the right side.

$$r^2mmr^3 = r^5 \qquad \text{for all figures.}$$
$$mr^2mr^3 = r \qquad \text{for all figures.}$$
$$mr^3mr^2 = r^{-1} \qquad \text{for all figures.}$$
$$rmrmrmrm = 1 \qquad \text{for all figures.}$$
$$r^3mr^2mr^{-3}mr = mr^3 \qquad \text{for all figures.}$$
$$mr^6mr^{-5}m = mr^3 \qquad \text{for the square.}$$
$$r^2mr^{-3}mr^2 = r^3 \qquad \text{for the square.}$$
$$r^2mr^{-3}mr^2 = r \qquad \text{for the triangle.}$$
$$mr^7mr^{-3} = r^2 \qquad \text{for the hexagon.}$$

The following Task was accidentally discovered by Nicolas Timbanidis as he was working through an early version of this book.

Task 3.2.15: For the square, the equation $mr^3mr^2m = mr$ is true. If you write both sides in reverse order then you get $mr^2mr^3m = rm$. It turns out that this equation is also true! Here is your Task: determine if writing both sides of a true equation in reverse order always results in another true equation.

Task 3.2.16: Let R stand for *some* rotation symmetry and let M stand for *some* mirror reflection symmetry of a finite figure. In the discussion following Task 3.2.4 it was noted that sometimes $RM = MR$ and sometimes $RM \neq MR$. Find a rule which tells you, for any particular M and R, whether or not RM equals MR.

3.3 Multiplication tables

The rules we found for combining symmetries are sufficient for computation purposes, but they don't give us the big picture. The collection of symmetries of a figure has itself an interesting structure. What exactly we mean by 'structure,' and what this structure is, will be explored fully in a later chapter. For now, we will just get a glimpse of this future topic. We will accomplish this by writing out the multiplication table for the symmetries of a figure. The table contains the information of how to combine any two symmetries of the figure, with the result written in our standard form.

Here is the multiplication table of the symmetries of the triangle:

$\triangle$	1	r	r^2	m	mr	mr^2
1	1	r	r^2	m	mr	mr^2
r	r	r^2	1	mr^2	m	mr
r^2	r^2	1	r	mr	mr^2	m
m	m	mr	mr^2	1	r	r^2
mr	mr	mr^2	m	r^2	1	r
mr^2	mr^2	m	mr	r	r^2	1

Each row and each column of the table is labeled with a symmetry written in our standard form, and the entries are also written in our standard form. The entries in the table show the result of multiplying the symmetry labeling the row by the symmetry labeling the column. We usually express this as (row)(column). It is important that the *row* goes on the *left* and the *column* goes on the *right*, and the result must be written in our standard form. For example, the 4th row is labeled by m, and the 3rd column is labeled by r^2, so the entry in the 4th row, 3rd column, is mr^2. Some of the entries take a bit more work. It seems like the entry in the 2nd row, 4th column should be rm. But rm is not in our standard form, so instead we use mr^2, because $rm = mr^2$, and mr^2 is in standard form.

Task 3.3.1: Write down the multiplication table for the symmetries of the square. Use your rules from the previous section to simplify the entries.

Task 3.3.2: Compare the multiplication tables for the square and the triangle, noting any patterns and properties they have in common. Make a guess at what the multiplication table for the pentagon would look like, and then write it down. Check a few entries in the pentagon table to see if it is correct.

3.4 Inverses

We originally described r^{-1} as the opposite rotation of r. A more precise definition would be that r^{-1} is the *inverse* of r, and we usually pronounce r^{-1} as "r inverse."

Definition. We say a is the **inverse** of b if $ab = 1$. In other words, doing b then doing a has the net effect of do–nothing.

In Task 3.2.5 we found that $r^{-1}r = 1$, so it is correct to say that r^{-1} is the inverse of r.

The multiplication table makes it easy to find the inverse of a symmetry. First locate the symmetry on the top row of the multiplication table. Then look down that column until you see a 1. The symmetry at the left end of that row is the inverse. Using this method for the triangle we find that r is the inverse of r^2, and mr is the inverse of mr. It is interesting that mr is its own inverse.

Task 3.4.1: Use the multiplication tables to find the inverses of the symmetries of the square and triangle.

Task 3.4.2: Can a symmetry have more than one inverse? What symmetries are their own inverses?

Task 3.4.3: What do you get when you take the inverse of the inverse of a symmetry? What does this say about the location of the 1's in a multiplication table?

3.5 The finite symmetry types

Every finite figure that we looked at had either only rotation symmetries, or else had an equal number of rotation and mirror reflection symmetries. Look back at our examples to check that this observation is correct.

If a finite figure has exactly N rotation symmetries and no mirror reflection symmetries, then we say the figure has **symmetry type** C_N. The C stands for "cyclic." For example, these figures have symmetry type C_3 :

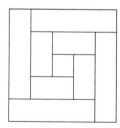

These have symmetry type C_4 :

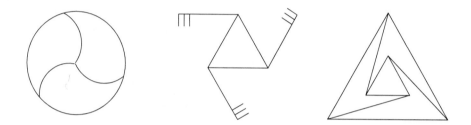

These have symmetry type C_2 :

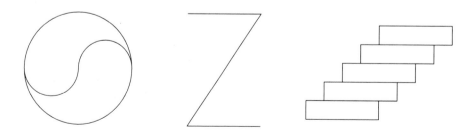

And these have symmetry type C_1 :

We usually think of a figure with symmetry type C_1 as having no symmetry, but it is more precise to say that it has do–nothing as its only symmetry.

If a finite figure has exactly N rotation symmetries and exactly N mirror reflection symmetries, then we say the figure has symmetry type D_N. The D stands for "dihedral." For example, these figures have symmetry type D_3 :

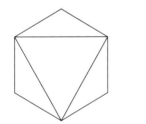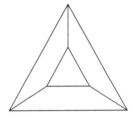

These have symmetry type D_4 :

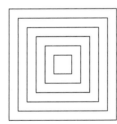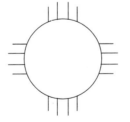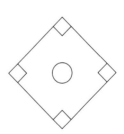

These have symmetry type D_2 :

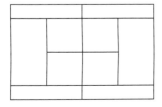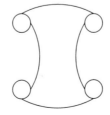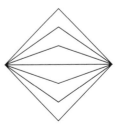

And these have symmetry type D_1 :

It is common to refer to a figure with symmetry type D_1 as having **bilateral symmetry**.

Task 3.5.1: Are there finite figures that don't have symmetry type C_N or D_N ? Either invent such a figure, or explain why none exists.

Task 3.5.2: Suppose that two finite figures are placed directly on top of one another. How does the symmetry type of the resulting figure compare to the symmetry types of the original figures? As a start, see what symmetry types you can produce by putting two figures with symmetry type D_3 on top of one another. Does it matter which two figures you start with?

3.6 Notes

Note 3.6.a: We defined a *finite figure* as one which has no nontrivial translation symmetries. There is nothing in the definition which requires the figure to 'look' finite. For example, a giant + formed by two perpendicular lines is a finite figure with symmetry type D_4, although the figure is infinitely large.

Note 3.6.b: A string of m's and r's, such as mr^2mr^3mrm, is often called a **word** in m and r.

Note 3.6.c: In this book we read a word in m and r starting from the right. Not all books do this. Which way you read a word is arbitrary, and Task 3.2.15 shows that the resulting equations are the same no matter which way you read, as long as you are consistent in your choice.

Note 3.6.d: We defined "a is the inverse of b" to mean $ab = 1$. It is standard to also require $ba = 1$. For the situations we encounter in this book, this second equation follows from the first, and vice versa. So when we say $a = b^{-1}$, you can think of $ab = 1$, or $ba = 1$, whichever you prefer.

Note 3.6.e: Our classification of finite figures into symmetry types C_N and D_N is not completely satisfactory. A better way to approach the topic would be to first define a notion of 'same symmetry type,' then find the possible symmetry types of finite figures, and then give appropriate names to the different symmetry types. That way, you could actually be sure that two figures with symmetry type C_4 actually were 'the same' in terms of symmetry. The precise definition requires the language of group theory, part of which will be introduced in Chapters 5 and 6. See Section 6.2 and the Notes at the end of Chapter 6 for more details.

4

Strip Patterns

4.1 Symmetries of strips. In the previous chapter we studied finite figures: shapes which have no translation symmetry. In this chapter we study **strip patterns**: patterns with translation symmetry in one direction. Here are some strip patterns.

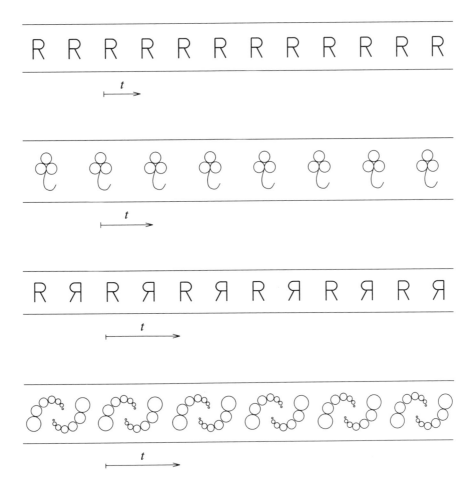

As above, we will always draw our strips so that the translation symmetries are in the left/right direction. To keep track of the translations, we adopt the following notation:

- t represents the smallest translation symmetry to the right.

We also let t^2 stand for doing t twice, t^{-1} for the inverse of t, and so on.

All the strips we consider will have translation symmetry, but they can have other symmetries also. When looking at symmetries of a strip, keep in mind that the strip is being treated as one rigid object. So a translation, rotation, reflection, or glide reflection applies to the entire strip, not just the small figures which make up the strip.

Some strips have rotation symmetry. The ○ marks the rotocenter:

R ○ ꓤ ○ R ○ ꓤ ○ R ○ ꓤ ○ R ○ ꓤ ○ R ○ ꓤ ○ R ○ ꓤ

□ ○ □ ○ □ ○ □ ○ □ ○ □ ○ □ ○ □ ○ □ ○ □

Task 4.1.1: True or False: A rotation symmetry of a strip must be a $\frac{1}{2}$–turn.

Strips can have mirror reflection symmetry. These strips have mirror reflection symmetry in a vertical mirror:

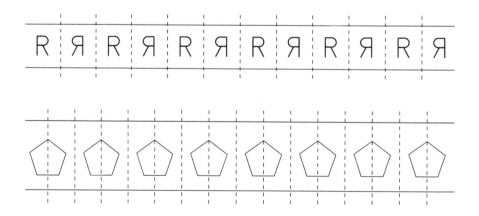

And these strips have mirror reflection symmetry in a horizontal mirror:

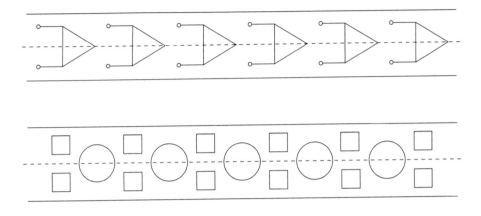

Task 4.1.2: True or False: If a strip has a mirror reflection symmetry, then the mirror line must be either horizontal or vertical: the mirror cannot be 'tilted.'

These strips have glide reflection symmetry.

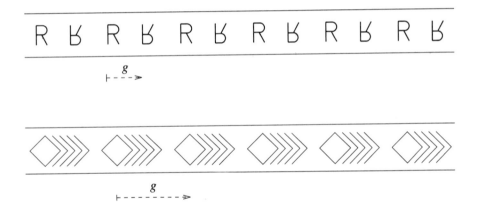

More notation:

- g represents the smallest glide reflection symmetry to the right.

The dotted arrow under the strip shows the amount of the glide.

Task 4.1.3: Draw some strips, and on each one label t, g, rotocenters, and mirror lines. Of course, not all strips will have all of these symmetries, so just label whatever symmetries occur.

Task 4.1.4: Let g^2 stand for doing g twice. Explain why g^2 is a translation. Experiment to see which translation it is. Note: the answer is not the same for all strips.

Task 4.1.5: Find some relationships between the amount of t, the amount of g, the distance between rotocenters, and the distance between vertical mirror lines.

4.2 Classifying strip patterns

In our study of finite patterns, we found that all finite patterns had one of two symmetry types: only rotation symmetries, or an equal number of rotation and reflection symmetries. This gave us a classification of the symmetry types of finite figures. In this section we classify the symmetry types of strip patterns.

A strip pattern, by definition, must have translation symmetry. It could also possibly have glide reflection, rotation, horizontal reflection, or vertical reflection symmetry. Not all combinations of these symmetries can occur. Finding exactly which combinations are possible is the point of this section.

Task 4.2.1: Draw some strips and list exactly which symmetries each one has. Try to find as many combinations as possible.

Task 4.2.2: List the combinations of symmetries you didn't find in Task 4.2.1.

Task 4.2.3: Propose some rules which explain why certain combinations of symmetries cannot exist by themselves. Here is an example rule: If there is a horizontal mirror and a vertical mirror then there also has to be a rotation symmetry.

Task 4.2.4: Either explain why the sample rule in Task 4.2.3 is true, or give a counterexample.

Task 4.2.5: Use your rules from Task 4.2.3 to explain why the combinations you listed in Task 4.2.1 cannot occur. Do your rules explain all the possibilities? If not, then either invent more rules, or find a strip with the missing symmetries.

Task 4.2.6: Explain why your rules from Task 4.2.3 are true. The material in Chapter 2 on combining rigid motions will be useful here.

Task 4.2.6 completes our classification of the symmetry types of strip patterns.

Task 4.2.7: The mathematician John Conway, from Princeton University, has given amusing names to the different symmetry types of strip patterns. His names are: hop; step; jump; sidle; spinning hop; spinning sidle, and spinning jump. The names go with the pattern of footprints you get when repeating each action. Draw the footprints corresponding to each name, and demonstrate the different walks to your friends.

4.3 Notes

Note 4.3.a: Inherent in our definition of a strip pattern is that it have a translation symmetry. A 'strip' of non–repeating figures does not qualify as a strip pattern. We have assumed, but never stated, that the translation symmetries must occur discretely. A horizontal line, which has translation symmetry by arbitrarily small amounts, does not qualify as a strip pattern.

Note 4.3.b: Our classification of strip patterns is not actually complete, because we have not shown that all strips with the same listed symmetries actually meet a reasonable definition of 'same symmetry type.' This shortcoming was also present in our classification of finite figures. See the Notes at the end of Chapter 6 for more details.

5

Wallpatterns

5.1 Rotation symmetry. In the previous chapter we studied strip patterns: patterns which have translation symmetry in one direction. In this chapter we study **wallpaper patterns**: patterns with translation symmetry in two different directions. The grids in Chapter 1 are all examples of wallpaper patterns. Here is another example:

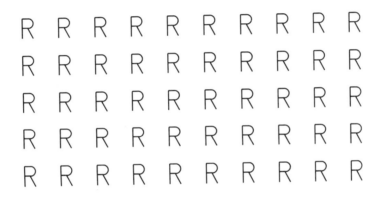

We see that the symmetries of this pattern consist only of translations: the pattern looks the same if we move it up/down and left/right by a whole number of units. In fact, the symmetries of this pattern are exactly the collection of "Today's Legal Moves" from the first part of Chapter 1. Wallpaper patterns with only translation symmetry are easy to understand; all the work has already been done in Chapter 1. Wallpaper patterns with other symmetries are more interesting. First we look at rotation symmetry.

To save typing, the author invents the following word:

wallpattern (wôl′păt′ern) *n.* short for wallpaper pattern.

We start building wallpatterns by stacking up copies of the following strip. The ∘ marks the rotocenters for the $\frac{1}{2}$-turn rotation symmetries of the strip:

ᴚ · R · ᴚ · R · ᴚ · R · ᴚ · R · ᴚ · R

43

Stacking the strips directly on top of one another we get this wallpattern:

 Я ∘ R ∘ я ∘ R ∘ я ∘ R ∘ я ∘ R ∘ я ∘ R
  ×   ×   ×   ×   ×   ×   ×   ×   ×
я ∘ R ∘ я ∘ R ∘ я ∘ R ∘ я ∘ R ∘ я ∘ R
  ×   ×   ×   ×   ×   ×   ×   ×   ×
я ∘ R ∘ я ∘ R ∘ я ∘ R ∘ я ∘ R ∘ я ∘ R
  ×   ×   ×   ×   ×   ×   ×   ×   ×
я ∘ R ∘ я ∘ R ∘ я ∘ R ∘ я ∘ R ∘ я ∘ R
  ×   ×   ×   ×   ×   ×   ×   ×   ×
я ∘ R ∘ я ∘ R ∘ я ∘ R ∘ я ∘ R ∘ я ∘ R

The ∘ from the strip pattern are also rotocenters for this wallpattern, and there are also other rotocenters, marked by × for 'extra.' These new rotocenters are also for $\frac{1}{2}$–turn rotation symmetries. You should take a moment to check that all the marked rotocenters are correct, and none is missing.

Is the appearance of these extra rotocenters surprising? Let's look at another example. Stacking the strips with a shift each time gives this wallpattern:

я ∘ R ∘ я ∘ R ∘ я ∘ R ∘ я ∘ R ∘ я ∘ R
×   ×   ×   ×   ×   ×   ×   ×   ×   ×
R ∘ я ∘ R ∘ я ∘ R ∘ я ∘ R ∘ я ∘ R ∘ я
×   ×   ×   ×   ×   ×   ×   ×   ×   ×
я ∘ R ∘ я ∘ R ∘ я ∘ R ∘ я ∘ R ∘ я ∘ R
×   ×   ×   ×   ×   ×   ×   ×   ×   ×
R ∘ я ∘ R ∘ я ∘ R ∘ я ∘ R ∘ я ∘ R ∘ я
×   ×   ×   ×   ×   ×   ×   ×   ×   ×
я ∘ R ∘ я ∘ R ∘ я ∘ R ∘ я ∘ R ∘ я ∘ R

Again we have the original ∘ rotocenters and extra × rotocenters, and again you should check that all of the rotocenters are correctly labeled.

Task 5.1.1: Suppose we stack the strips with just a small shift each time. Will we again get extra rotocenters, with the extra rotocenters shifted by just a small amount, or do the extra rotocenters only occur when we stack the strips so that the R's and я's lie in 'columns?'

Task 5.1.2: Why do the extra rotocenters appear?

Next we want to count the rotocenters in the wallpatterns we just looked at. At first this seems absurd, because obviously there are infinitely many of them. The idea is to use the symmetries to help distinguish the different rotocenters.

We say that two rotocenters in a wallpattern are **equivalent** if we can move one rotocenter to the other by a symmetry of the wallpattern. In the following example, the • rotocenters are all equivalent to each other, and no ∘ or × rotocenter is equivalent to any • rotocenter.

```
Я • R ∘ Я • R ∘ Я • R ∘ Я • R ∘ Я • R
  ×     ×     ×     ×     ×     ×     ×     ×     ×
Я • R ∘ Я • R ∘ Я • R ∘ Я • R ∘ Я • R
  ×     ×     ×     ×     ×     ×     ×     ×     ×
Я • R ∘ Я • R ∘ Я • R ∘ Я • R ∘ Я • R
  ×     ×     ×     ×     ×     ×     ×     ×     ×
Я • R ∘ Я • R ∘ Я • R ∘ Я • R ∘ Я • R
  ×     ×     ×     ×     ×     ×     ×     ×     ×
Я • R ∘ Я • R ∘ Я • R ∘ Я • R ∘ Я • R
```

It is important that you convince yourself that each • can be moved to any other •, and no • can be moved to any ∘ or ×. We sum this up by saying *the set of • rotocenters is an equivalence class.* There are four equivalence classes of rotocenters for the pattern:

```
Я • R ∘ Я • R ∘ Я • R ∘ Я • R ∘ Я • R
  ×   ⊗   ×   ⊗   ×   ⊗   ×   ⊗   ×
Я • R ∘ Я • R ∘ Я • R ∘ Я • R ∘ Я • R
  ×   ⊗   ×   ⊗   ×   ⊗   ×   ⊗   ×
Я • R ∘ Я • R ∘ Я • R ∘ Я • R ∘ Я • R
  ×   ⊗   ×   ⊗   ×   ⊗   ×   ⊗   ×
Я • R ∘ Я • R ∘ Я • R ∘ Я • R ∘ Я • R
  ×   ⊗   ×   ⊗   ×   ⊗   ×   ⊗   ×
Я • R ∘ Я • R ∘ Я • R ∘ Я • R ∘ Я • R
```

A better way of saying it is: this pattern has four inequivalent rotocenters. The idea of inequivalent rotocenters is applicable to strip patterns also. As you can check, the original strip has two inequivalent rotocenters, so it is correct to say that the wallpattern has twice as many rotocenters as the strip we used to build it.

Task 5.1.3: For several wallpatterns, label the rotocenters, and then figure out how many inequivalent rotocenters each has. Good examples to try are the grids from Chapter 1, and the examples following.

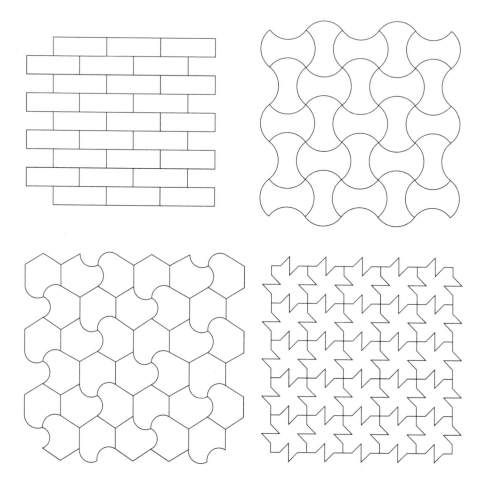

Each rotation symmetry has a rotocenter, and for each rotocenter we want to keep track of which rotations can occur. If the smallest rotation symmetry with that rotocenter is a $\frac{1}{2}$–turn, then we say the rotocenter has **order** 2; if the smallest rotation symmetry is $\frac{1}{3}$–turn, then we say the rotocenter has order 3; and so on. So instead of saying $\frac{1}{4}$–turn rotocenter, we say order 4 rotocenter. Think of it as a way to avoid using fractions.

Task 5.1.4: For the examples you did in Task 5.1.3, determine the order of each rotocenter.

Task 5.1.5: True or False: If two rotocenters are equivalent, then they have the same order. True or False: If two rotocenters have the same order then they are equivalent.

An obvious question is: What orders are possible for a rotocenter in a wallpattern? The answer is: 2, 3, 4, and 6. No other orders are possible. Is this surprising? An amusing exercise is to try and draw a wallpattern with order 5 rotocenters: you will not succeed. Although you can produce nice patterns with $\frac{1}{5}$–turn rotation symmetry, those patterns will not have translation symmetry. In fact, you cannot even draw a pattern which has two different order 5 rotocenters. A proof of this is given in the Notes at the end of the chapter.

5.2 Mirrors and glides

In this section we briefly discuss mirror symmetry and glide reflection symmetry in wallpatterns. These wallpatterns have mirror reflection symmetry:

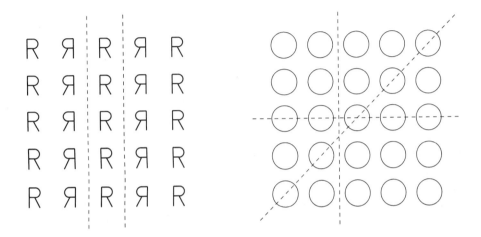

We have not drawn every mirror line for these patterns. Instead, we have only drawn representatives of the inequivalent mirror lines. We say that two mirror lines are **equivalent** if a symmetry of the pattern can move one mirror line to the other. This concept is identical to the one used in our discussion of rotocenters. You should take a moment to check that the mirror lines in the above examples have been drawn correctly.

Task 5.2.1: For several wallpatterns, label the inequivalent mirror lines. The examples from Task 5.1.3 are good patterns to try.

Task 5.2.2: Is there any connection between the rotocenters and the mirror lines of a wallpattern?

For strip patterns, we saw that a horizontal mirror line was automatically also a glide line. Similarly, any mirror line in a wallpattern must also be a glide line, so the only interesting glide lines are those which are *not* also mirror lines.

Task 5.2.3: Explain why a mirror line in a wallpattern must also be a glide line.

In the next examples we have labeled the inequivalent glide lines which are not also mirror lines. The first wallpattern has no mirror lines, but it has two inequivalent glide lines. The second has mirror lines, but it also has a glide line which is not a mirror line.

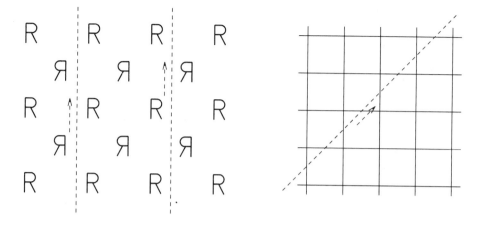

Task 5.2.4: For several wallpatterns, label the inequivalent glide lines which are not also mirror lines. Note: the hexagon grid and two of the patterns following Task 5.1.3 have interesting glide lines.

5.3 Classifying wallpatterns

We classified strip patterns by finding examples of which combinations of symmetries can occur, and then inventing rules to explain why other combinations are impossible. Given plenty of time and effort we could classify wallpatterns using the same method, but this would not be a productive use of our time. The result is that there are 17 different symmetry types of wallpatterns. Patterns with each symmetry type have been produced by various artists and artisans throughout history. A proof that there are only 17 possibilities was first given in 1891 by E.S. Fedorov. His paper was only published in Russian, and the proof first became generally known in 1924 when George Pòlya published his paper "Über die Analogie der Kristallsymmetrie in der Ebene." Pòlya's title reflects the fact that wallpatterns are related to the crystals studied by chemists. In his paper he gave an example pattern of each symmetry type, and these examples were to serve as inspiration for the Dutch artist M.C. Escher. At the end of this chapter are examples of all 17 symmetry types. Also at the end of the chapter is a flowchart based on one found in the excellent book *Symmetries of Culture* [SC] by Washburn and Crowe. This flowchart will enable you to determine the symmetry type of any wallpattern. The names given to each symmetry type are standard; you may find it amusing to try and figure out what the names actually mean.

Task 5.3.1: Use the flow chart to classify each of the sample wallpatterns. There are more than 17 patterns, so some symmetry types appear more than once. Do the patterns which are classified as 'the same' actually appear to be similar?

5.4 Basic units

The main topics of Chapter 1 were Legal Moves and basic units. In that chapter we only used translation symmetries. Now we return to those same topics, making use of the other types of symmetry.

Suppose we are given a collection of Legal Moves for a wallpattern. Recall that a **basic unit** is a region which covers the plane with no overlap as we apply all Legal Moves. It is important to keep in mind that the basic unit depends on the set of Legal Moves. If we change the set of Legal Moves, then the basic unit will also change.

The examples in this section are based on the square grid wallpattern:

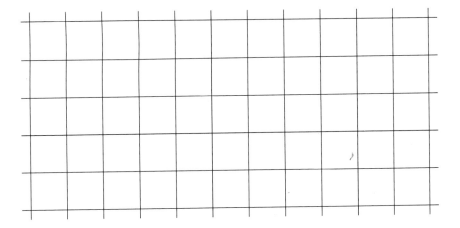

Our first set of Legal Moves is:

Legal Moves: The translation symmetries of the wallpattern.

This is exactly the situation discussed in Chapter 1, although the language used to describe it is different. A basic unit is one square. We know from Chapter 1 that there are many other choices of basic unit, but this choice seems natural.

Now we use a different set of Legal Moves:

Legal Moves: The translation and rotation symmetries of the wallpattern.

This time we have more Legal Moves, and a basic unit is $\frac{1}{4}$ of the square. It is worth spending some time checking that this shape covers the plane with no overlap when we use the translation and rotation symmetries of the square grid.

Our next example has even more Legal Moves:

Legal Moves: All symmetries of the wallpattern.

This time a basic unit is $\frac{1}{8}$ of the square. Again, check that you believe this is a correct basic unit for the set of all symmetries of the square grid.

Finally, we look at an example with very few Legal Moves:

Legal Moves: The left/right translation symmetries of the wallpattern.

This time we are not permitted to move up or down, or to rotate or reflect. For this set of Legal Moves we have a very large basic unit:

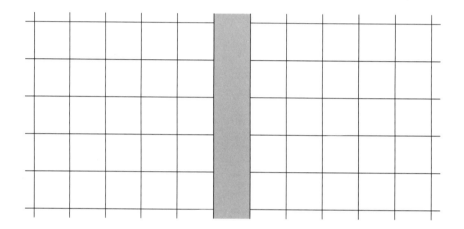

Task 5.4.1: Why does having more Legal Moves result in a smaller basic unit?

We will focus on the following collections of Legal Moves:

– All symmetries of the wallpattern.

– The translation and rotation symmetries of the wallpattern.

– Just the translation symmetries of the wallpattern.

Task 5.4.2: Explain how the idea of "basic unit" for the three sets of Legal Moves relates to the problem of printing wallpaper from a block of wood.

Task 5.4.3: For several wallpatterns, find a basic unit for each of the three collections of Legal Moves described above. The patterns from Task 5.1.3 are good examples to try.

Task 5.4.4: Invent a rule for how large the basic unit for the translation symmetries is, compared to the basic unit for the translation and rotation symmetries.

Task 5.4.5: Invent a rule for how large the basic unit for all symmetries is, compared to the basic unit for the translation and rotation symmetries.

5.5 Groups

If one chooses the Legal Moves unwisely, it may be impossible to find a basic unit. Here is an example using the square grid:

Bad Choice of Legal Moves: Translations by a whole number of units to the right.

This is not the same as the last example in the previous section, because now we are not allowed to translate to the left; we can only translate to the right. For this Bad Choice of Legal Moves it is impossible to find a basic unit. You should try to find one, and you will see that you need to use a very large shape to cover the plane, so large that it overlaps itself when you translate it to the right. If you make the shape smaller so that it doesn't overlap itself, it will not be able to cover the plane. In other words, there is no basic unit for this Bad Choice of Legal Moves.

Question: When does a set of Legal Moves have a basic unit?

Answer: When the set of Legal Moves is a *group*.

Definition. A set of Legal Moves is a **group** if:
1) Do–nothing is a Legal Move.
2) Any two Legal Moves in a row is another Legal Move.
3) The inverse of a Legal Move is also a Legal Move.

We see that our Bad Choice of Legal Moves was not a group because it violates condition 3). Translating right was Legal, but its inverse, translating left, was not Legal. It may be a good idea to look back in Section 3.4 to remind yourself about inverses.

The idea of a group is of fundamental importance to mathematics. Thousands of books and articles and lifetimes have been devoted to studying the theoretical aspects of groups. In fact, the topic is called Group Theory. Our study of groups will be confined to looking at examples, and we have actually seen several examples already.

Task 5.5.1: Check that the collections of Legal Moves in Section 5.4 all satisfy the three conditions for being a group.

The results of Task 5.5.1 are sufficiently important to state again. Each of the following collections of Legal Moves is a group:

– All symmetries of a wallpattern.

– The translation and rotation symmetries of a wallpattern.

– Just the translation symmetries of a wallpattern.

In other words, for each of the above collections of Legal Moves, we can find a basic unit. From now on we will speak of "the group of translation symmetries of a wallpattern," and so on.

Of course, not all collections of Legal Moves will be a group. We already saw an example of condition 3) being violated. This next choice of Legal Moves violates condition 2):

Bad Choice of Legal Moves: The rotation symmetries of the square grid.

Since a rotation of a rotation can sometimes be a translation, this choice does not satisfy condition 2).

And this choice of Legal Moves violates condition 1):

Bad Choice of Legal Moves: The reflection symmetries of the square grid.

Since do–nothing is not a reflection, this choice violates condition 1).

For more examples of things that *are* groups, turn to the next chapter.

5.6 Notes

Note 5.6.a: The result that 2, 3, 4, and 6, are the only possible orders of rotocenters for wallpatterns is called the *crystallographic restriction*. This name comes from the fact that crystals are commonly modeled on 3–dimensional grids.

Note 5.6.b: The following is an outline of the proof of the crystallographic restriction. The proof makes use of ideas from Task 2.5.5.

Suppose that there was a wallpattern with order 5 rotation symmetry. Choose one of the order 5 rotation symmetries and call it r_1, and label its rotocenter with ∘. Then find the order 5 rotation symmetry whose rotocenter is *closest* to ∘. Call the new rotation r_2, and label its rotocenter with •. We have this picture.

r_1 ∘   • r_2

Combining r_1 and r_2, we get $r_1 r_2$, which is yet another symmetry of the wallpattern.

Task 5.6.1: Show that $r_1 r_2$ is a $\frac{2}{5}$–turn rotation symmetry of the wallpattern, with rotocenter $\triangle$. Note that $\theta = 72° = \frac{1}{5}$–turn.

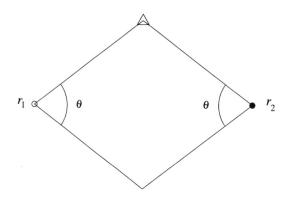

Task 5.6.2: Explain why $(r_1 r_2)^3$ is a $\frac{1}{5}$–turn rotation symmetry of the wallpattern, with rotocenter $\triangle$.

Task 5.6.3: Show that $\triangle$ is even closer to ∘ than • is.

Task 5.6.4: Conclude that our assumption that the wallpattern had an order 5 rotocenter was invalid, and thus it is impossible for any wallpattern to have order 5 symmetry. Conclude that it is impossible for any planar figure to have order 5 rotation symmetry at more than one rotocenter.

Task 5.6.5: Show that the above method is easily modified to prove that any odd number larger than 5 is impossible as the order of a rotation symmetry of a wallpattern. Conclude that any multiple of such a number is also impossible as the order of a rotation symmetry of a wallpattern.

Task 5.6.6: Modify the above method to show that 8, 12, 16, ..., are impossible as orders of a rotation symmetry of a wallpattern. Hint: start out the same, and then use ∘ and △.

That completes the proof of the crystallographic restriction.

Note 5.6.c: Even though $\frac{1}{5}$–turn rotation symmetry is not possible for wallpatterns, there are interesting shapes called *Penrose tiles* which fit together in a very close approximation to fivefold symmetry. See Martin Gardner's January 1977 column in Scientific American. That article is reprinted in his book *Penrose Tiles to Trapdoor Ciphers*, W.H. Freeman, 1989. Also see Chapter 7 of *The Mathematical Tourist: snapshots of modern mathematics*, by Ivars Peterson, W.H. Freeman, 1988. The best place to learn about these interesting tilings is at the Geometry Center at the University of Minnesota. Point your WWW browser to www.geom.umn.edu and look for *QuasiTiler*. Also see the program *Kali* to learn more about the 17 wallpatterns.

Note 5.6.d: The conditions in our definition of a group are often called:

1) Identity.

2) Closure.

3) Inverse.

The usual definition of a group also requires a condition called *associativity*:

4) $(ab)c = a(bc)$ for all a, b, and c in the group.

Our definition of *group* did not require this condition because we only consider groups of rigid motions, and for these, associativity is automatic. This is because our symmetries are functions, and function composition is associative.

Note 5.6.e: Given a set, it is not sensible to ask whether it is a group without first describing how the elements are to be combined. Since we are dealing with symmetries of figures, it is understood that we combine them in the same way that we combined rigid motions in Chapter 2. In some other settings, such as we will encounter in the next chapter, it is necessary to explicitly describe how things are to be combined.

Note 5.6.f: We have been looking at basic units for the symmetries of wallpatterns, but the idea works equally well for finite figures and strip patterns. Here is an example using the regular pentagon.

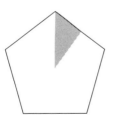

Basic unit for the group
of all symmetries.

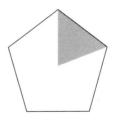

Basic unit for the group
of rotation symmetries.

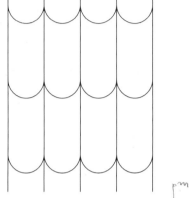

pm

cmm

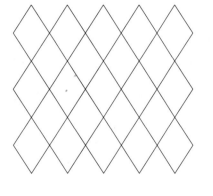

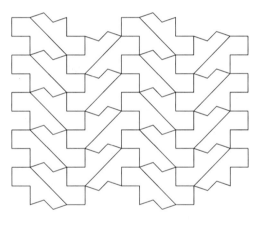

p6m

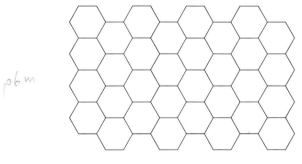

p4m

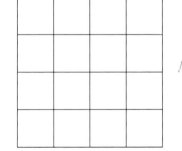

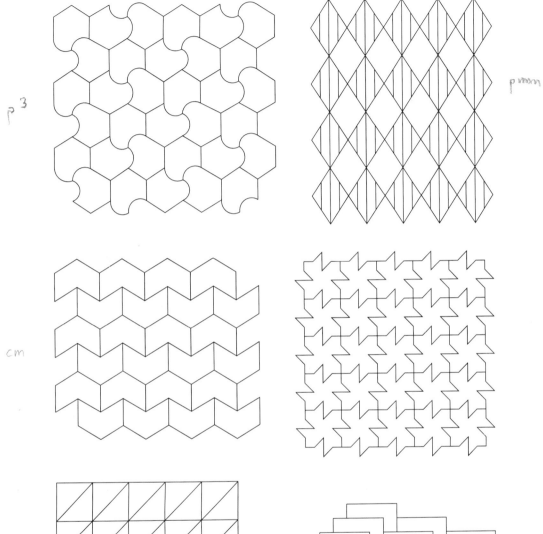

p 3

p mmm

cm

p l

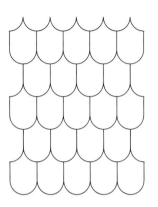

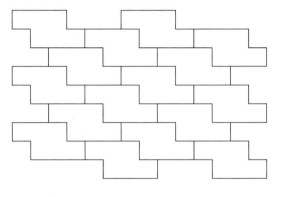

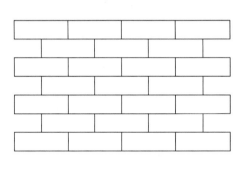

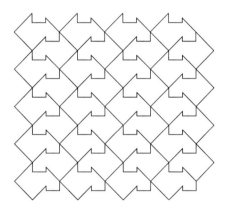

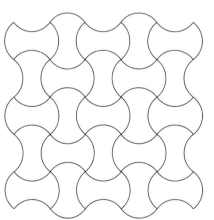

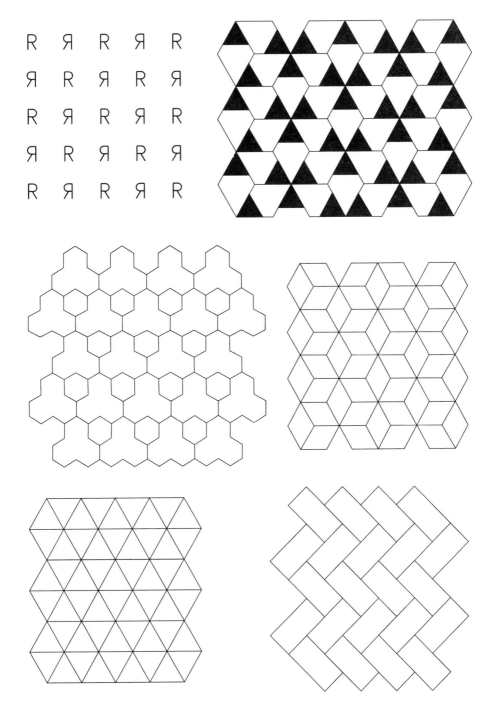

To classify a wallpattern, first find the smallest rotation symmetry, measured as a fraction of a full–turn.

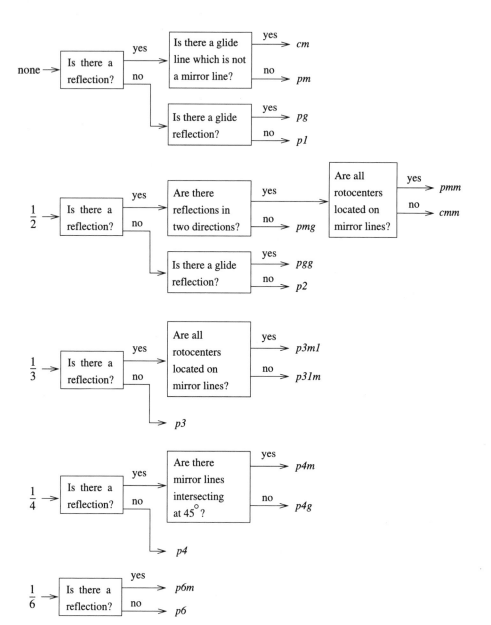

6

Finite Groups

6.1 Finite figures. Now we look for examples of groups among the finite figures. Our examples will be based on the square, but the ideas are applicable to all finite figures. Recall that the symmetries of the square can be represented by $\{1, r, r^2, r^3, m, mr, mr^2, mr^3\}$, with $r^4 = 1$. Do these symmetries form a group? That is, if the Legal Moves are the symmetries of the square, do these Legal Moves meet the definition of a group? Let's check the three conditions:

    1) Is do–nothing Legal? Yes, 1 is the same as do–nothing.

    2) If we combine two Legal Moves, do we get another Legal Move? Yes, in Chapter 3 we saw how to combine symmetries of the square.

    3) Is the inverse of a Legal Move also a Legal Move? Yes, in Task 3.4.1 we found the inverse of each symmetry of the square.

Therefore, the symmetries of the square are a group. Actually, most of the work above didn't make much use of the fact that we are dealing with the square.

Task 6.1.1: Explain why the collection of symmetries of *any* figure will always be a group.

    Next we look at the rotation symmetries of a square $\{1, r, r^2, r^3\}$, where $r^4 = 1$. Is this a group? Just check the conditions:

    1) Is do–nothing a rotation? Yes.

    2) Are two rotations in a row the same as one rotation? Yes.

    3) Is the inverse of a rotation also a rotation? Yes.

Therefore, the rotation symmetries of the square form a group.

    Given a collection of Legal Moves, we can always check the three conditions to determine if it is a group. This is a cumbersome method which gets tedious very fast. We now give an easier way to produce examples of groups. This method is based on Task 6.1.1.

    According to Task 6.1.1, given any figure, its symmetries form a group.

Here is an example figure:

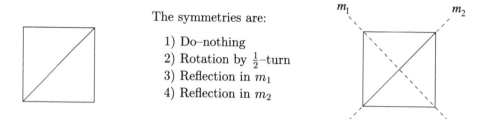

The symmetries are:

1) Do–nothing
2) Rotation by $\frac{1}{2}$–turn
3) Reflection in m_1
4) Reflection in m_2

By Task 6.1.1, the listed collection of four symmetries must form a group. Here is the trick: we can express those symmetries in terms of the symmetries of the square. Referring to your answers to Task 3.2.9, we find the correspondence is:

$$
\begin{array}{rcl}
\text{Do–nothing} & \longleftrightarrow & 1 \\
\text{Rotation by } \tfrac{1}{2}\text{–turn} & \longleftrightarrow & r^2 \\
\text{Reflection in } m_1 & \longleftrightarrow & mr^3 \\
\text{Reflection in } m_2 & \longleftrightarrow & mr
\end{array}
$$

Therefore, the set $\{1, r^2, mr, mr^3\}$, with $r^4 = 1$, is a group. Task 6.1.1 allows us to avoid explicitly checking the three conditions.

We can use the same method to show that our previous example $\{1, r, r^2, r^3\}$, with $r^4 = 1$, is a group. If we use this figure:

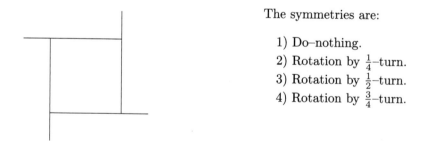

The symmetries are:

1) Do–nothing.
2) Rotation by $\frac{1}{4}$–turn.
3) Rotation by $\frac{1}{2}$–turn.
4) Rotation by $\frac{3}{4}$–turn.

Those four symmetries exactly correspond to $\{1, r, r^2, r^3\}$, with $r^4 = 1$, so this gives another way of checking that it is a group.

Task 6.1.2: Use this new method to find examples of groups. Some good figures to try are:

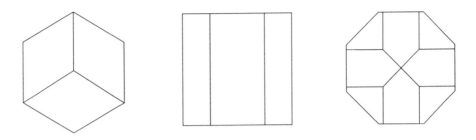

Think of the figures as 'starting from' the hexagon, square, and octagon, respectively. In each case you will need to specify $r^N = 1$ for some N.

Whenever we have a group we can write down its multiplication table. Here are the tables for the two examples shown previously.

⊓	1	r	r^2	r^3
1	1	r	r^2	r^3
r	r	r^2	r^3	1
r^2	r^2	r^3	1	r
r^3	r^3	1	r	r^2

◺	1	r^2	mr	mr^3
1	1	r^2	mr	mr^3
r^2	r^2	1	mr^3	mr
mr	mr	mr^3	1	r^2
mr^3	mr^3	mr	r^2	1

Task 6.1.3: Explain how each of the above multiplication tables can be thought of as 'part of' the multiplication table of the square.

The method we have been using to find examples of groups involves starting with one figure and its symmetry group, and then changing the figure slightly to get another figure with a smaller symmetry group. This smaller symmetry group is contained in the original symmetry group. A group contained within another group is called a **subgroup**. In the next section we look more closely at subgroups.

6.2 C_N and D_N, again

In Chapter 3 we classified the symmetry types of finite figures, finding that everything was of symmetry type C_N or D_N. Because we didn't yet know about groups, some of the statements in that chapter were incomplete.

The main correction is that C_N and D_N are actually symmetry *groups*, not just the names of symmetry types. For example, D_4 is the symmetry group of the square. Starting now, D_N will be called the dihedral group, and C_N will be called the cyclic group.

Recall that a group contained within another group is called a **subgroup**. You may have already noticed that C_N is a subgroup of D_N. For example, Figure 1 below has symmetry group C_4. In the previous section, we described its symmetries in terms of the symmetries of the square. Since the square has symmetry group D_4, this shows that C_4 is a subgroup of D_4. We previously noted that the multiplication table for C_4 looks like part of the multiplication table for D_4. Is this a coincidence? Of course not.

Figure 1   Figure 2

There are many other relations between these groups. For example, D_2 is a subgroup of D_4. This follows from what we have done with Figure 2 above. It has symmetry group D_2, and in the previous section we described its symmetries in terms of the symmetries of the square. Thus, D_2 is a subgroup of D_4.

There is some free information contained in the above analysis. Since C_2 is a subgroup of D_2, and D_2 is a subgroup of D_4, we conclude that C_2 is a subgroup of D_4.

The statement "D_2 is a subgroup of D_4" hides a subtle point. In Chapter 3 we wrote the elements of D_2 as $\{1, r, m, mr\}$, with $r^2 = 1$. In the previous section we wrote the elements of D_2 as $\{1, r^2, mr, mr^3\}$, with $r^4 = 1$. Both of these are descriptions of the same group. In fact, there is no 'best' way to write the elements of a group. We can choose different representations depending on what suits our present need. The important point is that while the description may change, the group stays the same.

Now let's look closer at the subgroups of C_N. We represent the elements of C_N by $\{1, r, r^2, \dots r^{N-1}\}$, with $r^N = 1$. There are two trivial cases:

Task 6.2.1: Explain why both $\{1\}$ and C_N are subgroups of C_N.

So we see that C_N always has at least two subgroups. For some values of N it will have other subgroups also.

Task 6.2.2: Check that $\{1, r^2\}$ is a subgroup of C_4. Check that both $\{1, r^3\}$ and $\{1, r^2, r^4\}$ are subgroups of C_6.

Task 6.2.3: Describe all of the subgroups of C_N.

Task 6.2.4: Describe all of the subgroups of D_N.

6.3 Addition

All of the groups we have seen so far are collections of motions. As a change of pace, we now look at some groups that are collections of numbers.

We will use a new form of addition to define these groups, and to make it clear that this is not the usual operation, we denote this new addition by $\oplus$.

Here is how it works:

$$a \oplus b = \text{the remainder when we divide } a + b \text{ by 7.}$$

For example, $3 \oplus 6 = 2$ because $3 + 6 = 9$, and when we divide 9 by 7 the remainder is 2. More examples are $5 \oplus 2 = 0$ and $1 \oplus 3 = 4$. This operation is called **addition modulo 7**, which we usually shorten to "addition mod 7."

This method of addition is something you use all the time. Addition mod 7 is how you tell the day of the week. For instance, in 7 days it will be the same day of the week as it is now, and 9 days from now is the same day of the week as 2 days from now.

Question: If today is Wednesday, what day will it be in 100 days?

Answer: If we divide 100 by 7 then the remainder is 2, so in 100 days it will be Friday.

We can do addition modulo any number, not just 7. For example, now let $\oplus$ stand for addition mod 12. The symbol is the same, but the new definition is

$$a \oplus b = \text{the remainder when we divide } a + b \text{ by 12.}$$

For instance, $8 \oplus 9 = 5$, $6 \oplus 6 = 0$, and $3 \oplus 5 = 8$. Counting mod 12 is how we tell time, and how we tell the month of the year. In 12 hours it will be the same time as it is now, if we ignore the am/pm, and 14 months from now it will be the same month as it is 2 months from now.

Question: Suppose it is 10:00 now. What time will it be in 100 hours?

Answer: If we divide 100 by 12 then the remainder is 4, so in 100 hours it will be 2:00.

If we want to keep track of whether 2:00 is day or night, then we would have to count mod 24.

Back to our study of groups. As an example we will use the numbers $\{0, 1, 2, 3\}$, and do addition mod 4. This is a group! To see this, just check the three conditions.

1) Is there a do–nothing? Yes, 0 is do–nothing. For example, $0 \oplus 3 = 3$, so 0 does nothing to 3. It is easy to see that 0 also does nothing to the other numbers.

2) If we combine two things in our group, do we get another? Yes, the operation $\oplus$ tells us to save the remainder when we divide by 4, and the possible remainders are 0, 1, 2, and 3.

3) Does everything have an inverse? Yes, we can check each case:

$$0 \oplus 0 = 0, \text{ so 0 is the inverse of 0}$$

$$3 \oplus 1 = 0, \text{ so 3 is the inverse of 1}$$

$$2 \oplus 2 = 0, \text{ so 2 is the inverse of 2}$$

$$1 \oplus 3 = 0, \text{ so 1 is the inverse of 3}$$

Therefore, we have a group. In exactly the same way, $\{0, 1, 2, 3, 4, 5\}$ is a group when we use $\oplus$ as addition mod 6, and so on.

Whenever we have a group we can write down a multiplication table.

Here is the multiplication
table for addition mod 4:

$\oplus$	0	1	2	3
0	0	1	2	3
1	1	2	3	0
2	2	3	0	1
3	3	0	1	2

Stare at that table for a while. We have seen this pattern before. This table looks just like the table for the rotations of the square at the end of Section 6.1. The only difference is that the earlier table had lots of r's everywhere; get rid of those r's and you end up with this table. What does this mean? It means that the group $\{0, 1, 2, 3\}$ with addition mod 4 is actually the same as the group of rotations of the square! Our attempt to deal with a group of numbers instead of a group of motions has failed miserably, for this group of numbers is actually the same as a group of motions.

More explanation is needed about what we mean by "the groups are *the same*." There are two parts to this. First, we must match up the elements in the two groups:

$$0 \longleftrightarrow 1$$
$$1 \longleftrightarrow r$$
$$2 \longleftrightarrow r^2$$
$$3 \longleftrightarrow r^3$$

Second, we must check that elements which are matched up behave in the same way. For example:

$$1 \oplus 2 = 3 \longleftrightarrow r\, r^2 = r^3$$
$$2 \oplus 3 = 1 \longleftrightarrow r^2\, r^3 = r$$
$$0 \oplus 2 = 2 \longleftrightarrow 1\, r^2 = r^2$$

These three pairs of equations are only a few of the many that need to be checked. Fortunately, we don't need to actually write out all possible equations. The multiplication tables contain all of the information about how the elements combine together, so by checking that the multiplication tables match up, we verify that corresponding elements behave the same.

Task 6.3.1: Write out the multiplication table for C_6, i.e. the rotations of the hexagon, and for addition mod 6, using the numbers $\{0, 1, 2, 3, 4, 5\}$, and verify that the multiplication tables match up.

Task 6.3.2: Invent a formula for inverses when $\oplus$ stands for addition mod N. What does this say about the placement of 0's in the multiplication table?

6.4 Multiplication

Next we use a new form of multiplication which is similar to the new kind of addition in the previous section. We write $\otimes$ for this new multiplication, and it is defined by

$$a \otimes b = \text{the remainder when we divide } a \times b \text{ by 5.}$$

For example, $3 \otimes 4 = 2$ because $3 \times 4 = 12$, and 12 divided by 5 leaves remainder 2. Similarly, $3 \otimes 2 = 1$ and $2 \otimes 2 = 4$. This operation is called "multiplication modulo 5," but we always simplify it to "multiplication mod 5." Using multiplication mod 5, the set $\{1, 2, 3, 4\}$ is a group. To see this, we need only check the three conditions.

1) Is there a do–nothing? Yes, 1 is do–nothing. For example, $1 \otimes 3 = 3$, so 1 does nothing to 3. It is easy to see that 1 also does nothing to the other numbers.

2) Do elements combine to give another element of the group? Yes, but a bit of explanation is needed. We multiply two numbers and then divide by 5, and we are not supposed to get 0 as the remainder, because 0 is not in the group. When dividing by 5, the only way to get 0 remainder is if you start with a multiple of 5. This can't happen in our case, because it is impossible to get a multiple of 5 by multiplying two of the numbers $\{1, 2, 3, 4\}$.

3) Does everything have an inverse? Yes, we can check each case:

$$1 \otimes 1 = 1, \text{ so 1 is the inverse of 1}$$
$$3 \otimes 2 = 1, \text{ so 3 is the inverse of 2}$$
$$2 \otimes 3 = 1, \text{ so 2 is the inverse of 3}$$
$$4 \otimes 4 = 1, \text{ so 4 is the inverse of 4}$$

Therefore, $\{1, 2, 3, 4\}$ is a group under the operation of multiplication mod 5. In the same way, the set $\{1, 2, 3, 4, 5, 6\}$ is a group under the operation of multiplication mod 7. Here are the multiplication tables for these groups:

$\otimes$	1	2	3	4
1	1	2	3	4
2	2	4	1	3
3	3	1	4	2
4	4	3	2	1

$\otimes$	1	2	3	4	5	6
1	1	2	3	4	5	6
2	2	4	6	1	3	5
3	3	6	2	5	1	4
4	4	1	5	2	6	3
5	5	3	1	6	4	2
6	6	5	4	3	2	1

Task 6.4.1: The multiplication tables for addition mod N and multiplication mod N appear more symmetric than the multiplication tables we wrote down in Section 3.3. Why is that so?

In Task 6.3.2 we found that when using addition mod 7, the inverse of the number a was $7 - a$. Finding inverses for multiplication mod 7 is not so easy. The multiplication table allows you to look up the inverse, but there is no simple formula for it.

Here are more examples of groups:

{1, 5, 7, 11} using multiplication mod 12

{1, 3, 5, 7} using multiplication mod 8

{1, 3, 5, 9, 11, 13} using multiplication mod 14

{1, 2, 4, 7, 8, 11, 13, 14} using multiplication mod 15

You may be wondering what weird method was used to choose those numbers. To figure out what is going on, you should first check that those examples actually are groups. Now, there is some work involved in doing that, but it can be done in a more interesting way than just tediously checking the three conditions. Here is a suggested method, applied to the first example above. It is easy to see that 1 is do–nothing, so two conditions remain to be checked.

First write out the multiplication table:

$\otimes$	1	5	7	11
1	1	5	7	11
5	5	1	11	7
7	7	11	1	5
11	11	7	5	1

Second, use the information contained in the multiplication table to check the remaining conditions. To check that the elements combine to give other elements in the group, you need only verify that the entries inside the table consist only of numbers from the top row and first column. To see if everything has an inverse, you need only check that the number 1, which is do–nothing for this group, appears exactly once in each row and column. Since both of these are true for the above table, we have a group.

Task 6.4.2: Use the above method to check that the other three examples are also groups.

Task 6.4.3: Experiment to see that changing the numbers in the above examples destroys the fact that we have a group. In particular, the given numbers were not chosen haphazardly.

Task 6.4.4: What numbers would you use to make a group for multiplication mod 10? For multiplication mod 18? Check that your answer is correct by writing out the multiplication table.

It seems that the groups in this section are just groups of numbers, not groups of motions. But actually, these are groups of motions too! This is the topic of the next section.

6.5 Rearrangements

In this section we use multiplication tables to show that all groups are groups of motions.

If you examine the multiplication tables we have written down, you will see that the following are true:

- Each element in the group appears exactly once in each row.

- Each row is different.

Those statements are also true if you replace "row" by "column."

Task 6.5.1: Explain why the statements are true.

Another way of phrasing those observations is, "each row is a different rearrangement of the group." Now, each row is labeled by a group element, so *each element tells you how to rearrange the group.* In other words, each element is a motion; it is the motion of rearranging the group! This is amazing. Any group is automatically a group of motions.

A motion of rearrangement isn't quite the same as a rigid motion of the plane, but this doesn't tarnish the interesting fact that all groups can be thought of as groups of motions. The observation that all groups are actually groups of rearrangements was first made in the mid–19th century by the mathematician Arthur Cayley. This result is commonly called "Cayley's Theorem."

In the next section we explore rearrangements in more detail.

6.6 Permutations

The traditional word for rearrangement is **permutation**. We will use both words in this section.

Our first examples will be based on rearranging the numbers 1 2 3 4 5 6. One way to describe a permutation is to show where each number goes. Here are two example permutations which we have named f and g:

$$f: \begin{array}{cccccc} 1 & 2 & 3 & 4 & 5 & 6 \\ \downarrow & \downarrow & \downarrow & \downarrow & \downarrow & \downarrow \\ 3 & 5 & 4 & 6 & 2 & 1 \end{array} \qquad g: \begin{array}{cccccc} 1 & 2 & 3 & 4 & 5 & 6 \\ \downarrow & \downarrow & \downarrow & \downarrow & \downarrow & \downarrow \\ 2 & 6 & 5 & 3 & 1 & 4 \end{array}$$

For instance, f takes the list 1 2 3 4 5 6 and rearranges it to get 3 5 4 6 2 1. We can also think of f as rearranging any list of 6 things. It takes the 1st element and puts it 3rd, it takes the 2nd element and puts it 5th, and so on.

We are interested in studying groups, so it is good news that *the collection of all permutations of* 1 2 3 4 5 6 *is a group.* Both f and g are elements of that group. To see that we have a group, we must check the three conditions.

1) Is there a do–nothing? Yes, do–nothing is the rearrangement which just leaves everything where it currently is.

2) Does combining two permutations give another permutation? Yes, for if we rearrange, and then rearrange again, that is the same as one big rearrangement. As an example, we can combine our sample permutations f and g to get a new permutation fg. Recall that fg means first do g then do f:

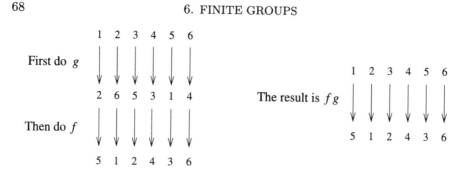

3) Does every permutation have an inverse? Yes, just "unrearrange." For example, this permutation is the inverse of f:

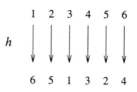

To show that h is the inverse of f, we must check that hf is the same as do–nothing.

So it is correct that h is the inverse of f.

Since the set of all permutations of 1 2 3 4 5 6 satisfies the three conditions, we conclude that it is a group. This group is called S_6, pronounced "ess six." The S stands for 'symmetric.' Similarly, the group of permutations of 1 2 3 4 5 is called S_5, and so on.

The notation we have been using to describe a permutation is very cumbersome. We now introduce a better method called **cycle notation**. Here is an example cycle: (5 2 1 3). This cycle is interpreted as "5 goes to 2, 2 goes to 1, 1 goes to 3, and 3 goes to 5." We read left to right, and when we reach the end we go back to the beginning. If a cycle doesn't mention a number, then the number does not move. For instance, the cycle just given doesn't affect the number 4. Since the numbers not mentioned in a cycle do not move, the "empty cycle" () is do–nothing. A cycle with just one number, such as (3), is also do–nothing. This is because 3 goes to 3, so 3 doesn't move, and nothing else is mentioned, so nothing else moves either.

To combine cycles, we start on the right. For example, (3 5 2)(6 3 1) means "6 goes to 3, 3 goes to 1, and 1 goes to 6, then 3 goes to 5, 5 goes to 2, and 2 goes to 3." Note that 3 was moved more than once.

Warning. When combining cycles, we first do the cycle on the *right*. But within each cycle, we read left to right.

If we have to deal with several cycles, it is easiest if each number appears in only one cycle. We call this **disjoint cycles.** For example, (2 3 5)(7 4 6 1) is a product of disjoint cycles. The usefulness of disjoint cycles is that each number is only being moved once. An important fact is that any permutation can be written as a product of disjoint cycles. For example:

$$
\begin{array}{ccc}
\begin{array}{cccccc}
1 & 2 & 3 & 4 & 5 & 6 \\
\downarrow & \downarrow & \downarrow & \downarrow & \downarrow & \downarrow \\
3 & 5 & 4 & 6 & 2 & 1
\end{array}
& \text{is the same as} & (2\ 5)(1\ 3\ 4\ 6)
\end{array}
$$

with f labeling the left diagram.

$$
\begin{array}{ccc}
\begin{array}{cccccc}
1 & 2 & 3 & 4 & 5 & 6 \\
\downarrow & \downarrow & \downarrow & \downarrow & \downarrow & \downarrow \\
2 & 6 & 5 & 3 & 1 & 4
\end{array}
& \text{is the same as} & (1\ 2\ 6\ 4\ 3\ 5)
\end{array}
$$

with g labeling the left diagram.

$$
\begin{array}{ccc}
\begin{array}{cccccc}
1 & 2 & 3 & 4 & 5 & 6 \\
\downarrow & \downarrow & \downarrow & \downarrow & \downarrow & \downarrow \\
5 & 1 & 2 & 4 & 3 & 6
\end{array}
& \text{is the same as} & (6)(4)(1\ 5\ 3\ 2)
\end{array}
$$

with fg labeling the left diagram.

We could also represent fg as (1 5 3 2) because both (4) and (6) are do–nothing. The cycle representations were found by tracing through the permutation, finding in succession where each number goes.

Task 6.6.1: Show that the permutations in S_3 can be represented by $\{(), (1\ 2), (2\ 3), (3\ 1), (1\ 2\ 3), (1\ 3\ 2)\}$.

If we have a product which is not disjoint, then we can turn it into a product of disjoint cycles. For example, (3 5 2 1)(6 2 1) = (6 1)(2 3 5). This answer was found by just tracing through the cycles. Here are the steps:

First pick any number. We will start with 6.   (6

First 6 goes to 2, then 2 goes to 1, so the net result
is 6 goes to 1.   (6 1

Now we find where 1 goes. First 1 goes to 6, then
6 doesn't move, so the net result is 1 goes to 6.   (6 1)
This gives one cycle.

Now pick another number. We pick 2.   (6 1)(2

First 2 goes to 1, then 1 goes to 3, so the net result   (6 1)(2 3
is 2 goes to 3.

First 3 stays where it is, then it goes to 5, so the   (6 1)(2 3 5
total effect is 3 goes to 5.

First 5 stays where it is, then it goes to 2, so the
total effect is 5 goes to 2. This completes another   (6 1)(2 3 5)
cycle. All numbers have been used, so we are done.

This method can be used to turn any product of cycles into a product of disjoint
cycles. A useful exercise is to multiply the cycle representations given above for
f and g to see that you get the cycle representation given for fg.

Task 6.6.2: Write down a product of cycles and use the above method to turn
it into a product of disjoint cycles.

Task 6.6.3: What is the inverse of a cycle?

Task 6.6.4: Write down the multiplication table for S_3. Use the elements of S_3
listed in Task 6.6.1.

Task 6.6.5: Explain why a product of disjoint cycles can be rearranged without
changing the resulting permutation. For example, (1 2 3)(4 5 6 7) is the same
as (4 5 6 7)(1 2 3).

Task 6.6.6: Give an example to show that if you rearrange non–disjoint cycles
then the result can be a different permutation.

To complete the circle of ideas in this section, we show how cycle notation
can be used to describe the symmetries of a square. Here are two symmetries:

We can describe these symmetries by showing how each one rearranges the labels
we put on the square: $m =$(c d)(b a) and $r =$(a b c d). Any symmetry of any
finite figure can be described in this way.

Task 6.6.7: Using the labeling above, determine which symmetries of the square correspond to (a c)(b d) and to (a c).

Task 6.6.8: How many elements are in the group S_n ? In other words, how many ways are there to rearrange the numbers 1 2 3 . . . n ?

A cycle containing two elements, such as (3 5), is called a **2–cycle**, or a **transposition**. It switches two elements, and everything else is unchanged. It is possible to write any permutation as a product of 2–cycles, although the 2–cycles may not be disjoint. For example, (1 2 3)=(2 3)(1 3).

Task 6.6.9: Find another expression for (1 2 3) as a product of 2-cycles.

Task 6.6.10: Write (1 2 3 4) as a product of 2-cycles.

6.7 Notes

Note 6.7.a: The technical term for two groups being 'the same' is *isomorphic*. The matching up of the elements in the two groups is called an *isomorphism*. Those words come from the Greek roots *iso*, meaning 'same,' and *morph*, meaning 'form.' To repeat the definition from Section 6.3, an **isomorphism** between two groups is a matching of the elements of the groups which also gives a matching of the multiplication tables of the groups. If such a matching exists, then we say the two groups are **isomorphic**.

Task 6.7.1: Show that S_3 is isomorphic to D_3. Note: you have already written down the multiplication tables for both groups, so once you decide on a matching, it will be easy to check if that matching works.

One way to see an answer to Task 6.7.1 is to label the corners of a triangle, and then see how the symmetries of the triangle rearrange those labels. It is just a coincidence that S_3 is the same as D_3. For larger values of N, the groups S_N and D_N are different. However, by labeling the corners of a regular N–gon we see that D_N is a subgroup of S_N for $N \geq 4$.

Note 6.7.b: In Chapter 3 we classified finite figures into symmetry types C_N and D_N. At that time we did not actually explain the way in which figures having the same symmetry type are actually 'the same.' Now we can give the official definition.

Definition. Two finite planar figures have the **same symmetry type** if there is an isomorphism between their symmetry groups such that the isomorphism matches rotations with rotations and reflections with reflections.

The above definition does *not* merely say that the symmetry groups of the two figures must be isomorphic; there is an additional condition that there be an isomorphism which matches reflections with reflections and rotations with rotations. This distinction is relevant, as demonstrated by the following Task.

Task 6.7.2: Show that C_2 is isomorphic to D_1. Explain why C_2 and D_1 are different symmetry types.

To get the definition of *symmetry type* for strip patterns and wallpatterns, just add the condition that translations must be matched with translations, and glide reflections must be matched with glide reflections.

7

Cayley Diagrams

7.1 Generators. In this chapter we examine a way to look at a group as one entire unit, as opposed to just a collection of separate elements. We will accomplish this by studying the Cayley diagram of the group.

In Chapter 3 we saw that the symmetry group of the triangle can be represented by $\{1, r, r^2, m, mr, mr^2\}$. For our present purposes, the key observation is that everything is built from r and m. Since r and m can be combined to produce all of the other elements, we say that r and m **generate** the group.

Knowing that r and m are generators of the group does not explain the exact way they combine to give the other elements. One method of keeping track of how these combinations occur is called a **Cayley diagram**. Below is the Cayley diagram for the symmetry group of the triangle, and following it is an explanation of what the different parts of the diagram represent.

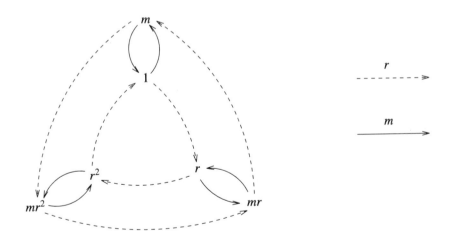

The diagram has two parts: each element in the group appears once, and each generator has an arrow associated to it. Here is how the arrows work: Let's consider the element mr^2. If we multiply by m we get $m(mr^2) = r^2$. In other words, m 'moves' mr^2 to r^2. Therefore we have an m arrow ⟶ going from mr^2 to r^2. Similarly, $r(mr^2) = mr$, so we have an r arrow ┄┄┄> from mr^2 to mr. Each element will have one m arrow and one r arrow coming out of it. To find where that arrow goes, we multiply on the left and simplify. You can check that all of the arrows have been drawn correctly.

Task 7.1.1: Use the Cayley diagram to show $rmr^2mrmr^2 = mr^2$.

7.2 Rearranging basic units

Having seen the Cayley diagram for the symmetry group of the triangle, it is fairly easy to guess the Cayley diagram of the square. We are going to write down the Cayley diagram of the symmetry group of the square, but we will use a method which is also useful for finding the Cayley diagrams of the symmetry groups of other figures. The method makes use of the basic units for the group of symmetries. Basic units for the symmetry groups of finite figures are mentioned in Note 5.6.f at the end of Chapter 5.

Step 1. Divide the square into basic units for the group of all symmetries. Label each basic unit.

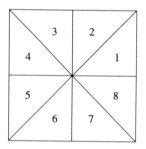

Step 2. Make little diagrams showing how m and r rearrange the basic units. We will still use ———→ for m and - - - - -→ for r. Recall that m stands for reflection in a vertical mirror.

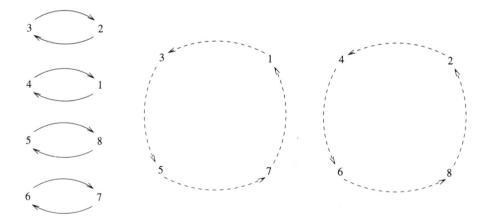

Step 3. Imagine that the little diagrams are assembly instructions. Glue all of the 1's together, and all of the 2's together, and so on. Here

is a partial gluing for the square:

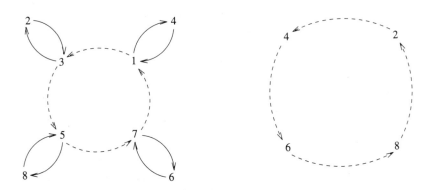

To finish, we must glue those two pieces together. This can be done if we first flip over one piece. Here is the result:

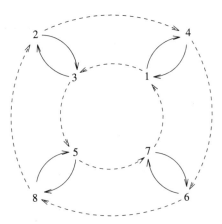

This is the Cayley diagram for the symmetry group of the square.

Having seen these two examples it is an easy matter to write down the Cayley diagram of any finite figure.

It might seem bothersome that the diagram is labeled with basic units, instead of group elements. However, there is a simple remedy. To convert the labels from basic units to group elements, first replace any basic unit label by '1,' the do–nothing element of the group. Then follow the arrows in the diagram and replace the other basic unit labels with the appropriate group element. This is the idea behind Task 7.1.1.

7.3 Strip patterns

Next we find the Cayley diagram for the symmetry group of some strip patterns. We will use the same three steps that we used to find the Cayley diagram of the square.

One difficulty faces us immediately, and that is finding generators for the group of symmetries of the strip. More precisely, the problem is finding a 'good' set of generators. Since we will have to glue together little diagrams from each generator, we want to use as few generators as possible. Also, we want to choose generators so that the resulting little diagrams are easy to fit together. Our plan is this: first find a generating set and form all of the little diagrams. Then use the smallest number of generators necessary to give a completely connected diagram. If the chosen pieces do not easily fit together, then replace one of the chosen generators with an unused generator. Repeat this process until a nice Cayley diagram appears.

The set of generators we will start with is the inequivalent rotations and inequivalent mirror reflections, along with the smallest translation and the smallest glide reflection symmetry. You should think about why this is a good set to start with.

Here is an example strip:

This strip has a translation symmetry t, and two inequivalent rotation symmetries r_1 and r_2, with rotocenters $\circ$ and $\bullet$, respectively. We have shaded a basic unit. Now divide the strip into basic units and label each one, and choose a different arrow for each generator. When you try these on your own, you should use different colored pens for the different arrows.

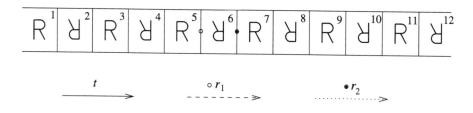

Here are the little diagrams for each generator:

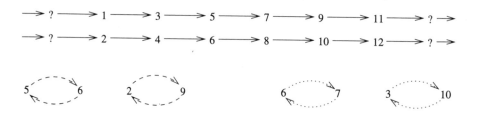

The ? represents various unnamed basic units. Now we start gluing things together. We will be finished when we have a connected diagram in which every number appears exactly once. Let's start by choosing generators t and $\circ r_1$. Here is a partial gluing:

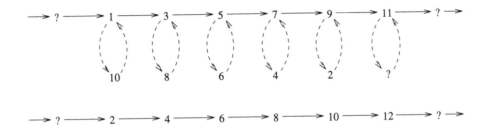

Every number appears above, so we will be done if we can glue those two pieces together. To do this we must flip over one of the pieces. We end up with the following Cayley diagram of the symmetry group of the strip:

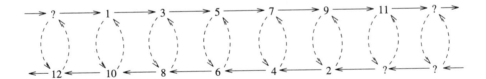

That is not the only possible answer. If we use t and $\bullet r_2$ as generators then we get essentially the same picture, but if we use $\circ r_1$ and $\bullet r_2$ then we get this very different Cayley diagram:

Comparing those two diagrams, we see that the way t and $\bullet r_1$ combine to produce all symmetries of the strip is very different from the way that $\circ r_1$ and $\bullet r_2$ combine to give the other symmetries.

Here is another strip:

This time we have t translation, g glide reflection, and m mirror reflection in a horizontal mirror. A basic unit is shaded. Here is a labeling of the basic units, and an arrow for each generator:

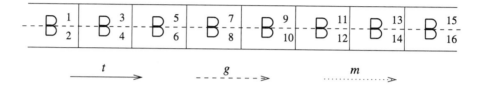

And here are the little diagrams:

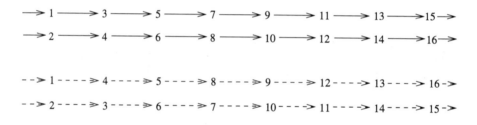

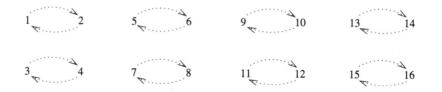

Gluing t and m together gives this Cayley diagram:

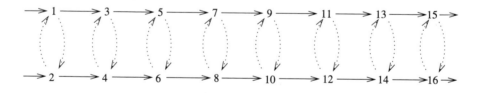

Note that this is *not* the same as the Cayley diagram from the previous strip. Gluing g and m together gives a similar picture.

We can also glue t and g together, but it is impossible to draw it without having the arrows cross:

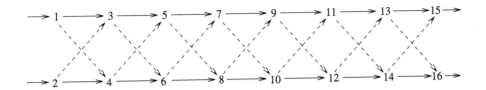

This diagram is not of interest to us: we want to find Cayley diagrams without crossing arrows. If this had been the result of our first attempt with this strip, then we would discard either t or g and use m instead. This process of replacing a generator when the little diagrams don't fit together properly is of major importance when doing a more complicated example.

Task 7.3.1: Find a Cayley diagram of the group of symmetries of this strip:

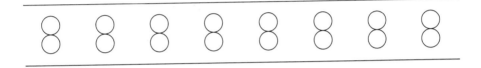

7.4 Wallpatterns

Next we find the Cayley diagram of the symmetry group of this wallpattern:

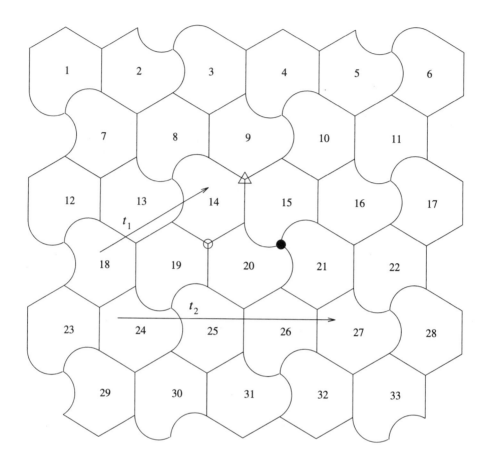

We labeled the basic units, two translation symmetries t_1 and t_2, and the three inequivalent rotocenters $\circ r_1$, $\bullet r_2$, and $\triangle r_3$. This accounts for all symmetries of the wallpattern. To find the Cayley diagram, we follow the plan used in the last section. First we write down little diagrams for each generator.

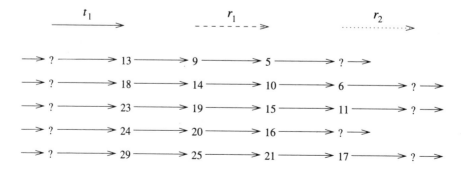

$\longrightarrow ? \longrightarrow 30 \longrightarrow 26 \longrightarrow 22 \longrightarrow ? \longrightarrow$

$\longrightarrow ? \longrightarrow 31 \longrightarrow 27 \longrightarrow ? \longrightarrow$

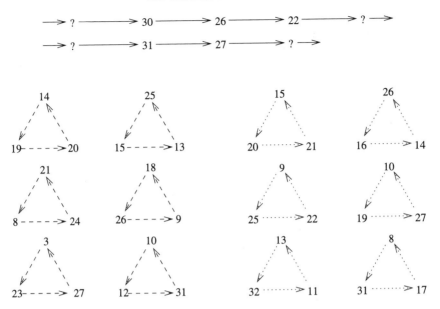

To save space, we have not drawn the little diagrams for t_2 or r_3, but we can do so later if we need them. Now we begin gluing things together. At this stage, we don't know how many generators we will need, but we hope that two will be sufficient. For no good reason, let's start with t_1 and r_1.

Beginning with this piece from t_1,

$\longrightarrow ? \longrightarrow 23 \longrightarrow 19 \longrightarrow 15 \longrightarrow 11 \longrightarrow ? \longrightarrow$

we glue on pieces from r_1 to get,

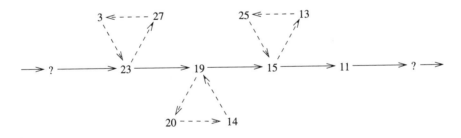

The exact way that the various pieces are oriented is arbitrary, and we may need to rearrange them later. Next we glue on another piece from t_1. A reasonable choice is this one,

$\longrightarrow ? \longrightarrow 18 \longrightarrow 14 \longrightarrow 10 \longrightarrow 6 \longrightarrow ? \longrightarrow$

because the two 14s will connect and the resulting figure will not be too large.

This is the result:

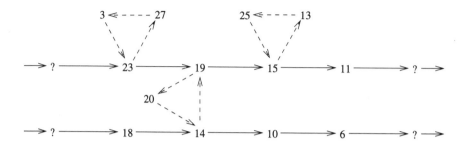

Now there are two pieces from r_1 which can be connected:

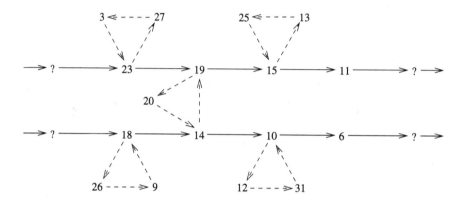

Things are not looking very pretty, but there is still hope that it will all come out right. We notice that the following two pieces from t_1 can both be connected to the above diagram in two places:

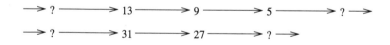

If you examine where those two pieces connect to the partially built diagram, you will see that gluing both of them will give a horribly tangled mess. It is time to abandon one of our generators and choose another. Since the long strings from t_1 were found to be unwieldy, we will throw away t_1 and use r_2 instead. So now we start over and try to build the Cayley diagram using r_1 and r_2.

Here is a piece from r_1:

Gluing on pieces from
r_2, we get this:

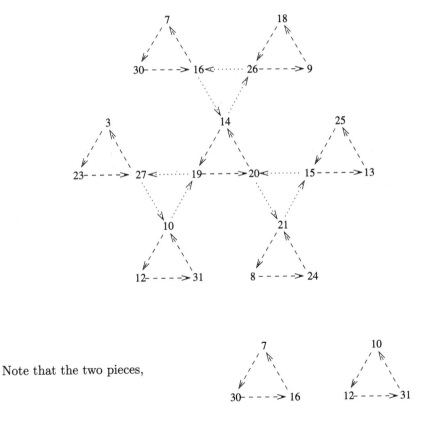

The exact way in which the new pieces are arranged can be changed later, if
necessary. Now glue on more pieces from r_1:

Note that the two pieces,

were not on our previous list of little diagrams from r_1, but they would have
been, if that list was complete. An important point: you can always construct
more little diagrams. Occasionally this requires labeling more basic units.

Now attach pieces from r_2:

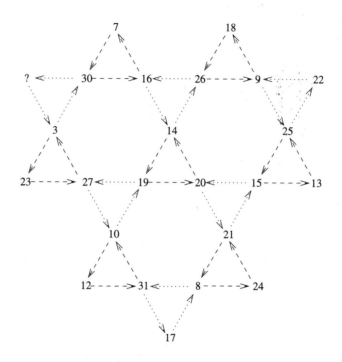

Those pieces fit perfectly, and now we see the pattern. The Cayley diagram is a collection of interlocked six–pointed stars. Drawing everything on a smaller scale, we get the following Cayley diagram for the symmetry group of the wallpattern:

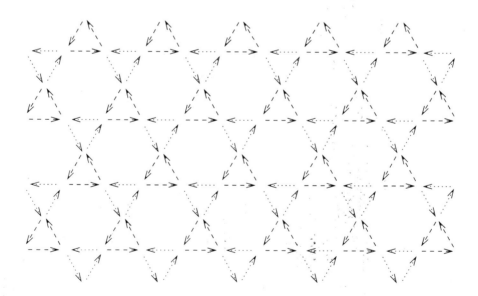

Task 7.4.1: Find Cayley diagrams for the symmetry groups of the following wallpatterns.

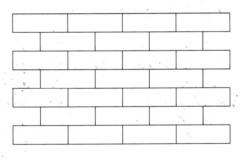

8

Symmetry in the Real World

8.1 Analyzing patterns. This chapter discusses the practical uses of Groups and Symmetry, and gives a list of interesting topics for further study. An excellent book on analyzing patterns is *Symmetries of Culture* [SC] by Washburn and Crowe. The book *Symmetry* [S] by Weyl has lots of nice ideas and pictures, although it sometimes uses complicated words to explain simple concepts. The most comprehensive source on 2–dimensional symmetry is *Handbook of Regular Patterns* [HRP], by Stevens.

The first thing to do when analyzing a pattern is to decide which category it falls under:

Finite figures: no translation symmetries.

Strip patterns: translation symmetry in one direction.

Wallpatterns: translation symmetry in two different directions.

This classification is not as simple as it seems, because many patterns are made of various smaller patterns.

Suppose we want to analyze the print on this pillowcase:

The only symmetry of the entire pillowcase is do–nothing. To get an interesting pattern to analyze, we must concentrate on one part of the pillowcase.

The flowers form a wallpattern:

This wallpattern can be analyzed using the methods of Chapter 5. Also, each row or column of flowers can be analyzed as a strip pattern, and each individual flower can be analyzed as a finite figure.

The shape along the right border is a finite figure with symmetry type D_2. However, if we concentrate on the inner part of that figure, we can imagine it as a strip pattern:

This process of taking a small figure and imagining it as a part of a larger pattern is something we do all the time. Since any physical object must necessarily be finite, the only way to get an infinitely repeating pattern is to use your imagination. This is the exact process we used when saying that the flowers on the pillowcase form a wallpattern.

The interplay between the various patterns can be studied for both its mathematical and its artistic content. For example, two strip patterns can be placed together to form a larger strip pattern. In some cases, the combined strip will have more symmetry than either of the component strips, and in other cases the combined strip will be less symmetric. If you find such an example, say, on a piece of Indian pottery, then a discussion of the artistic interaction of the various strips is just as important as a discussion of the mathematical interaction.

8.2 Patterns in art and architecture

The mathematics in this book can be used to analyze a variety of patterns which appear in art and architecture. Below are some suggestions on writing a paper on the mathematics of real–world patterns. Following the suggestions is a list of possible topics. The list includes only a few of the many places one can find interesting patterns.

The following are good things to put in your paper:

Give background information on your topic.

Explain the mathematical terms you will use. Simple diagrams are a great help in explaining the various concepts.

Make a drawing or photocopy of the patterns, and label rotocenters, mirror lines, basic units, and so on. Use the flow–chart to classify the pattern.

Discuss the problems you encountered while analyzing the patterns. For example, were the objects you studied made up of several different patterns combined together? Did the patterns show small variations which had to be ignored in order to have more symmetries? If so, were these small variations intentional?

Do all of the patterns you examined fall into just a few symmetry types? If so, can you explain why? Do you think that you found all of the possible examples?

The topic is of general interest, so you want to make sure that your paper is comprehensible to a variety of readers. Explain the various terms you use. Indicate where you found the various patterns, so that your paper can be used as a guidebook.

Bricks. Analyze the patterns of bricks used on the outside of buildings. Investigate the patterns on your campus, or in your neighborhood, or in the business section of your town. Compare the classification bricklayers use for the different patterns to the classification we use in Groups and Symmetry.

Decorative floors and walls. Pick a particular place and analyze the patterns you find there. Many museums and churches have a wealth of interesting patterns to analyze. You could analyze and compare the patterns appearing in various restaurants in your town.

The art of M.C. Escher. Escher constructed many fascinating wall-patterns. See the book *Visions of Symmetry* [VS] for a complete catalog of his 'regular divisions of the plane.'

The art of William Morris. Morris was the leading wallpaper designer of the Victorian era. The book *The Art of William Morris* [AWM] is a good source, and any good library will have books on his work. When analyzing wallpaper, it is interesting to relate the idea of a basic unit for the various groups of symmetries to the practical problem of printing actual wallpaper.

William Morris and M.C. Escher. It is interesting to compare the work of these artists, particularly in terms of the symmetry of their designs.

Islamic art. There is some debate as to how many different wallpatterns occur in the Alhambra.

African weavings. Kente cloth, produced by the Ashanti of West Africa, has an interesting history. Yoruba *adire* cloth, produced in Nigeria, has many fascinating patterns to analyze. These textiles have an important place in the culture of the people who produce it.

Indian pottery. There are many interesting patterns to be found on the pottery of native American peoples. Navaho rugs is also a fascinating topic.

Rugs and carpets. So–called "Oriental rugs" are produced in many countries. Each has its distinctive features, and in many cases this is related to the symmetry of the rug's design.

Amish quilts. Traditional handmade Amish quilts have a beautiful geometric simplicity to them. Close examination of the stitching patterns reveals a more elaborate structure. It is interesting to compare the mathematical idea of 'basic unit' to the practical problem of piecing together bits of fabric to form a quilt.

8.3 Mathematical projects

The projects in this section build upon the mathematics we studied in the book.

The 15–Puzzle. The 15–puzzle can be analyzed using group theory. The 15–puzzle consists of a 4×4 array containing the numbers 1 to 15, along with one blank space. The object is to scramble the numbers by sliding them around, and then unscramble them so that the numbers are back in order. It is a surprising fact that if you pry out two pieces and switch them, then it becomes impossible to solve the puzzle. The idea of this project is to use group theory to explain why this is true. Here are the steps:

1. A **transposition** is a cycle with two elements, such as $(a\ b)$. A transposition switches two elements, and leaves all the others unchanged. Show that any permutation can be written as a product of transpositions.

2. A permutation can be expressed as a product of transpositions in several different ways. Show that for any given permutation, the ways it can be expressed as a product of transpositions either all have an even number of transpositions, or all have an odd number of transpositions. We say that the permutation is **even** or **odd**, respectively.

3. Show that the set of even permutations on n letters is a group. This group is called the **alternating** group, and it is denoted by A_n. The alternating group A_n is a subgroup of the symmetric group S_n.

4. Explain how the legal moves in the 15–puzzle correspond to permutations. Then show that these moves correspond to even permutations. That is, the legal moves are in the alternating group.

5. Conclude that after illegally switching two numbers, the puzzle cannot be solved.

6. Compute how many legal positions there are in the 15–puzzle.

The main part is Step 2. It may seem 'obvious' that the product of an even number of transpositions cannot equal the product of an odd number of transpositions, but actually *proving* it takes some work.

The great puzzle inventor Sam Lloyd offered a large cash prize to anyone who could solve the 15–puzzle starting from the solved position with the numbers 14 and 15 switched. As you have shown in this project, his money was completely safe.

More arithmetic mod N**.** This project is about 2×2 matrices. You don't need any experience with these; everything is explained below.

Here is an example 2×2 matrix: $\begin{pmatrix} 2 & 5 \\ 4 & 8 \end{pmatrix}$. We multiply matrices using this rule:

$$\begin{pmatrix} a & b \\ c & d \end{pmatrix} \begin{pmatrix} A & B \\ C & D \end{pmatrix} = \begin{pmatrix} aA + bC & aB + bD \\ cA + dC & cB + dD \end{pmatrix}$$

The usual way to think about it is, rows from the matrix on the left are multiplied by columns from the matrix on the right. Some examples:

$$\begin{pmatrix} 2 & 4 \\ 3 & 7 \end{pmatrix} \begin{pmatrix} 5 & 1 \\ 2 & 4 \end{pmatrix} = \begin{pmatrix} 18 & 18 \\ 29 & 31 \end{pmatrix}$$

$$\begin{pmatrix} 6 & 2 \\ -1 & 3 \end{pmatrix} \begin{pmatrix} 2 & -3 \\ 1 & 1 \end{pmatrix} = \begin{pmatrix} 14 & -16 \\ 1 & 6 \end{pmatrix}$$

$$\begin{pmatrix} 3 & 1 \\ 5 & 2 \end{pmatrix} \begin{pmatrix} 2 & -1 \\ -5 & 3 \end{pmatrix} = \begin{pmatrix} 1 & 0 \\ 0 & 1 \end{pmatrix}$$

Check those examples to see that they are correct.

Now we add a wrinkle: we will do our arithmetic mod 7. The above examples now become:

$$\begin{pmatrix} 2 & 4 \\ 3 & 0 \end{pmatrix} \begin{pmatrix} 5 & 1 \\ 2 & 4 \end{pmatrix} = \begin{pmatrix} 4 & 4 \\ 1 & 3 \end{pmatrix}$$

$$\begin{pmatrix} 6 & 2 \\ 6 & 3 \end{pmatrix} \begin{pmatrix} 2 & 4 \\ 1 & 1 \end{pmatrix} = \begin{pmatrix} 0 & 5 \\ 1 & 6 \end{pmatrix}$$

$$\begin{pmatrix} 3 & 1 \\ 5 & 2 \end{pmatrix} \begin{pmatrix} 2 & 6 \\ 2 & 3 \end{pmatrix} = \begin{pmatrix} 1 & 0 \\ 0 & 1 \end{pmatrix}$$

Again, check that these examples are correct. In each case, we did the addition and multiplication mod 7, and we replaced each number by the number from 0, 1, 2, 3, 4, 5, 6, which represents it mod 7.

If we did arithmetic mod 3 then we would have the following examples:

$$\begin{pmatrix} 2 & 1 \\ 0 & 1 \end{pmatrix} \begin{pmatrix} 2 & 1 \\ 2 & 1 \end{pmatrix} = \begin{pmatrix} 0 & 0 \\ 2 & 1 \end{pmatrix}$$

$$\begin{pmatrix} 0 & 2 \\ 2 & 0 \end{pmatrix} \begin{pmatrix} 2 & 0 \\ 1 & 1 \end{pmatrix} = \begin{pmatrix} 2 & 2 \\ 1 & 0 \end{pmatrix}$$

$$\begin{pmatrix} 0 & 1 \\ 2 & 2 \end{pmatrix} \begin{pmatrix} 2 & 2 \\ 1 & 0 \end{pmatrix} = \begin{pmatrix} 1 & 0 \\ 0 & 1 \end{pmatrix}$$

Again, check.

The plan is to turn the 2×2 matrices into a group. We will do arithmetic mod N, using the numbers $0, 1, 2, \ldots, N-1$. As you experiment, it might be good to start out using the numbers $0, 1, 2$, and do arithmetic mod 3, or use the numbers 0 and 1 and do arithmetic mod 2. Using our definition of matrix multiplication, the product of 2×2 matrices is always a 2×2 matrix, so we must concentrate on the problem of identity and inverse.

1. Find the matrix which is do–nothing. It is common to represent this by the letter I, and it is called the **identity matrix**. By definition, it has the property that for any matrix g, we have $Ig = g$ and $gI = g$.

2. Recall that "h is the inverse of g" means $hg = I$. Or equivalently, $gh = I$. Determine which matrices have an inverse. Explain how to find the inverse.

Part 2 is the main content of this project. By experimenting, you will see that not all matrices have an inverse. Finding a method of describing which matrices have an inverse is the key step in describing which matrices should be in your group.

3. Show that, when $N = 2$, the group you get is the same as D_3. That is, find a way to match up the matrices with the elements $\{1, r, r^2, m, mr, mr^2\}$ of D_3, so that the multiplication tables also match up. Note: for other values of N it is not possible to match up the matrix group with a group we have seen previously.

4. Determine how many matrices are in your group. Note: this is tricky, and you should be satisfied if you can do it for a few specific values of N.

Generators, relations, and Cayley diagrams. A common way to describe a group is in terms of *generators* and *relations*. This is called a **presentation** of the group. For example,

$$D_4 = \langle r, m \mid r^4 = 1, m^2 = 1, rm = mr^{-1} \rangle,$$

is a presentation of the symmetry group of the square. We say that r and m are the **generators**: the group is built from them. We call $r^4 = 1$, $m^2 = 1$, and $rm = mr^{-1}$, the **relations**: they completely describe how the generators combine to produce the elements of the group. Here is another example:

$$G = \langle a, b \mid ab = ba \rangle$$

This group is the same as the group of Today's Legal Moves from the first part of Chapter 1. Equivalently, this is the symmetry group of the first wallpattern shown in Chapter 5. If you think of a and b as 'up one' and 'right one,' then the relation $ab = ba$ just says 'up one and right one is the same as right one and up one.' Here is the Cayley diagram of the group G, using $a \longrightarrow$, and $b \dashrightarrow$ for the generators:

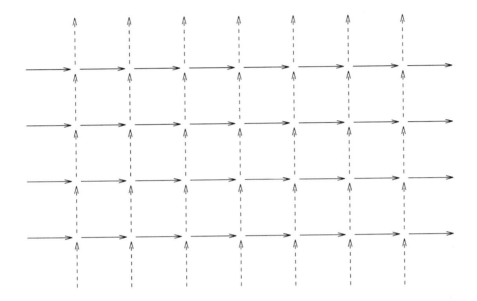

Here is the point: The Cayley diagram tells you how to find a presentation for the group. So if you had the diagram, you could then write down the presentation. This is explained below.

To find the presentation, we need generators and relations. The generators are just sitting there in the diagram. To find the relations, we look at the ways to walk a closed path in the diagram. Each different closed path gives us a relation. A path is **closed** if it ends at the place it began. The way it works is: *closed path of generators* = 1. The dotted line shows a closed path in the previous Cayley diagram:

This closed path tells us that $aba^{-1}b^{-1} = 1$. We read the generators in the path in order, using the generator if the path is in the same direction as the arrow, and its inverse if the path is in the opposite direction of the arrow. Note that this relation can be rearranged to say $ab = ba$, which is the relation we expected. A bit of checking will convince you that the other closed paths in that Cayley diagram all give the same relation.

It is a good exercise to check that in the Cayley diagram of the square, there are three different closed paths, and these give the three relations we have been using. The relations might not initially look like our standard form, but a bit of rearrangement will make them look familiar.

1. Use the Cayley diagrams from Chapter 7 to find presentations for the symmetry groups of some strip patterns and wallpatterns.

2. Explain how to use the Cayley diagram to reduce long "words" in the generators, such as $ab^3a^2b^{-4}a^3b^5a^{-2}$, to a simpler form. In particular, describe how to tell if a word is the same as do–nothing. See Task 7.1.1 for a similar problem.

One way to produce a new group from an old one is to put in more relations. For example, here is a group:

$$H = \langle a, b \mid ab = ba, a^2 = 1, b^2 = 1 \rangle$$

We see that this is the group G with two more relations added. We will use the Cayley diagram for G to find the Cayley diagram for H.

Here is the Cayley diagram for G, with all the elements labeled:

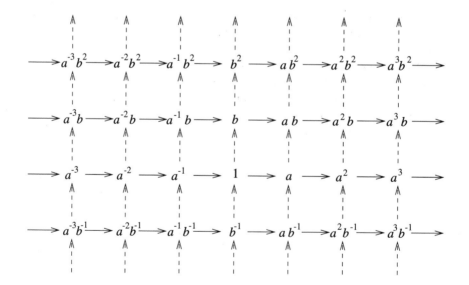

The relation $a^2 = 1$ implies that $a^3 = a$, $a^4 = 1$, $a^5 = a$, and so on, and $a^{-1} = a$, $a^{-2} = 1$, and so on. The same rules hold for b. Using these rules to relabel the

Cayley diagram, we get this pattern:

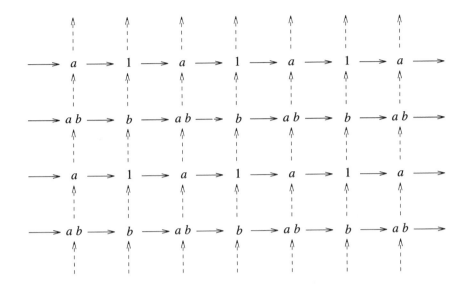

Now, to find the Cayley diagram for H, we just glue all the 1's together, glue all the a's together, and so on.

3. Do the gluing described above to find the Cayley diagram for H.

4. Add different relations to G and see what the resulting Cayley diagram looks like. Some good ones to try: $a^2 = 1$ and $b^3 = 1$, or $a^3 = 1$ and $b^3 = 1$. Try some more until you notice a pattern.

3–dimensional symmetry. There are five regular solids: tetrahedron, cube, octahedron, dodecahedron, and icosahedron. Each of these shapes has lots of symmetry. A good reference is *Shapes, Space, and Symmetry* [SSS].

1. Find all symmetries of each regular solid.

Building cardboard models is a good way to understand and describe the symmetries.

2. How do the symmetry groups of these solids relate to each other? For example, it is possible to fit a tetrahedron inside a cube, so that the corners match up. What does this say about the relation between the symmetry group of the square and the symmetry group of the tetrahedron? Find other interesting ways that one regular solid can be put inside another.

3. Which groups C_N and D_N are contained in a symmetry group of a regular solid?

4. If you put a dot in the center of each face of a regular solid, and connect the dots of adjacent faces, then you get another regular solid. This new solid is called the **dual** of the original solid. This example shows that the octahedron is the dual of the cube:

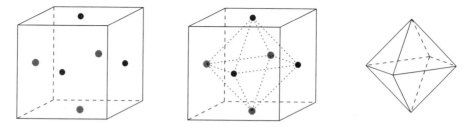

Find the dual of each regular solid. What relationships exist between a solid and its dual? In particular, what is the relationship between the symmetries of a regular solid and the symmetries of its dual?

5. In the plane, we found four kinds of rigid motions. What are the possible rigid motions of 3–dimensional space?

Magic square wallpatterns. Here are two magic squares.

2	9	4
7	5	3
6	1	8

1	15	14	4
12	6	7	9
8	10	11	5
13	3	2	16

The one on the left is a 3×3 magic square, and the one on the right is a 4×4 magic square. In a magic square, the sum of the numbers in any row, column, or diagonal is the same.

– Find ways to use rearrange one magic square to get other magic squares.

Take a magic square and use it to tile the plane. The resulting wallpattern will have only translation symmetry.

This wallpattern was made from the 3×3 square shown above.

```
2 9 4 2 9 4 2 9 4 2 9 4
7 5 3 7 5 3 7 5 3 7 5 3
6 1 8 6 1 8 6 1 8 6 1 8
2 9 4 2 9 4 2 9 4 2 9 4
7 5 3 7 5 3 7 5 3 7 5 3
6 1 8 6 1 8 6 1 8 6 1 8
2 9 4 2 9 4 2 9 4 2 9 4
7 5 3 7 5 3 7 5 3 7 5 3
6 1 8 6 1 8 6 1 8 6 1 8
```

You should imagine that each number is inside a little square. Now color the odd numbered squares white and the even numbered squares black. Does the resulting wallpattern have more than just translation symmetry? Investigate to see how much or how little symmetry the colored magic square wallpattern can have. Does it matter whether the original magic square was 3×3, or 4×4, or larger? What if you color each numbered square with red, white, or blue, depending on whether the number gives remainder 0, 1, or 2 when divided by 3? Here are some references:

Chapter 4 of *Mathematics on Vacation* by Joseph S. Madachy, Scribners, 1966.

"Magic Cubes and the 3–adic Zeta Function." Allan Adler, *The Mathematical Intelligencer* **14**, No 3, Summer 1992, p14.

Magic Squares and Cubes by W.S. Andrews, Dover Books.

8.4 Random projects

These projects don't fit under the headings of the previous two sections.

Kinship structures. Many so–called primitive tribes have intricate rules of kinship, descent, and inheritance. In many cases, these rules are related to the groups we have studied in this book. An excellent account of one such group can be found in Chapter 3 of *Ethnomathematics* [E], by Ascher. The main example in that chapter is a native Australian tribe which bases its kinship relations on D_4, the symmetry group of the square.

1. In *Ethnomathematics* the symmetries of the square are written differently than the representations we use: she writes the reflection on the right, instead of on the left. Reanalyze the kinship rules of the Warlpiri so that the result looks like our representation of the symmetry group of the square.

2. Find other sources which describe similar kinship structures, and analyze them to see what group is being used.

You will probably find it necessary to introduce the terminology which anthropologists use to describe kin relations.

Chemistry. It is said that one can use symmetry to predict the boiling point of alkanes. Is this true? If it is, how do you determine which molecule is 'more symmetric'? How does this describe what is occurring on the molecular level?

Crystals provide another interesting topic of study, although much of the subject is not easily accessible to nonexperts.

Tiling a wall. A homeowner wants to put new tile on the kitchen wall. There are two types of square tile available: plain and fancy. The plain tiles are inexpensive and are one solid color. The fancy tiles are expensive and have a decorative design on them. The homeowner plans to buy several boxes of identical plain tile and one or two boxes of identical fancy tile. The tiles will be placed in a square grid on the wall. Your job is to advise the homeowner of the various ways the two tiles can be combined to make interesting wallpatterns. Some questions to answer:

1. Which of the 17 wallpatterns can be made with the given tiles?

2. What effect does the symmetry of the fancy tile have on the wallpaper possibilities?

3. What ratio of plain to fancy gives the best choice of wallpatterns?

Try to make your paper useful to a typical homeowner interested in tiling the kitchen wall. Include any interesting wallpatterns you find.

Now suppose that an eccentric homeowner wants to use non–square tiles. Furthermore, the homeowner wants to tile different walls with wallpatterns having different symmetry types, but all of the patterns will be made with the same tile. Cost is no object: the homeowner is willing to pay to have special tiles made. Your job is to design a tile which can be used to make several different wallpatterns.

Make your own patterns. Devise repeating patterns similar to those created by Escher. Your project should include your finished pictures, and also a detailed description of the process you went through to create each picture. If you keep all of your scratchwork, it can be pasted together to show your progress. Then you can add comments to describe the thought processes you went through at each step. The *Handbook of Regular Patterns* [HRP] has some suggestions on how to create repeating patterns, but those suggestions seem to be of limited use. Carefully examining Escher's work is more likely to be useful. You will find that creating a good repeating Escher–style pattern takes quite a bit of time and ingenuity. A record of the steps you went through will help other people understand the process.

There are computer programs which automate the process of producing wallpatterns with a given symmetry type. One such program is *Kali*. It is available free from the Geometry Center at the University of Minnesota. You can get it by anonymous ftp from geom.umn.edu. A version for the Macintosh can be found in the directory /pub/software/Kali.

Bibliography

[AC] *Africa Counts*, by Claudia Zaslavsky, Lawrence Hill Books, 1990.
Interesting account of the mathematics of the native African peoples.
Section 5, "Pattern and Shape," is relevant to this book.

[AWM] *The Art of William Morris*, by Aymer Vallance, Dover, 1988.
Description of his life and work. 40 color plates of his wallpatterns.

[C] *Connections*, by Jay Kappraff, McGraw Hill, 1990.
An interesting book which touches on many of the topics discussed in
the present book.

[E] *Ethnomathematics: A Multicultural View of Mathematical Ideas*, by Marcia Ascher, Brooks/Cole, 1991.
Presents an interesting account of the mathematical sophistication of
'primitive' people. The chapter, "The Logic of Kin Relations" is fascinating. "Symmetric Strip Decorations" is a nice introduction.

[EAS] *M.C. Escher: Art and Science*, H.S.M. Coxeter, ed., North–Holland, 1986.
A collection of 35 papers on mathematical symmetry. Most relate to
Escher's work, and most have nice pictures. A good place to see how
others have analyzed real–world symmetry. Some of the papers are
very mathematical.

[FS] *Fivefold Symmetry*, István Hargittai, ed., World Scientific, 1992.
Various papers on pentagonal symmetry. The paper, "800–Year–Old
Pentagonal Tiling..." suggests that Penrose tilings were invented in
12th–century Iran.

[FSE] *Fantasy and Symmetry: The Periodic Drawings of M.C. Escher*, by Caroline
MacGillavery, Harry N. Abrams, 1976.

[GCIA] *Geometric Concepts in Islamic Art*, by I. El–Said and A. Parman, World of
Islam Festival, 1976.
Mathematics and Islamic Art.

[HRP] *Handbook of Regular Patterns*, by Peter Stevens, MIT Press, 1981.
A comprehensive book on regular patterns. Millions of examples, and
some reasonable guides on how to construct interesting patterns.

[KS] *Knots and Surfaces: a guide to discovering mathematics*, by David Farmer and Theodore B. Stanford, American Mathematical Society, 1996.
A book in the same style as *Groups and Symmetry*.

[MS] *Mathematical Snapshots*, by Hugo Steinhaus, various publishers.
Nice chapters on a variety of mathematics, written for a general audience. All of it is interesting, and two or three of the chapters are relevant to this book.

[S] *Symmetry*, by Hermann Weyl, Princeton University Press, 1989.
Interesting essays on real–world symmetry.

[SC] *Symmetries of Culture*, by Dorothy Washburn and Donald Crowe, University of Washington Press, 1988.
An excellent book. Designed for anthropologists who want to analyze patterns. Lots of details. Many pictures. Useful flow–charts to classify patterns. Plenty of references.

[SSS] *Shapes, Space, and Symmetry*, by Alan Holden, Dover, 1991.
Pictures of hundreds of 3–dimensional symmetric shapes. Lots of regular and semi–regular solids. Discussion of their symmetries.

[SS2] *Symmetry* and *Symmetry 2*, István Hargittai, ed., Pergamon Press, 1986, 1989.
Two huge books of papers on all aspects of symmetry.

[TP] *Tilings and Patterns*, by Branko Grünbaum and G.C. Shephard, Freeman, 1987.
Very mathematical and not that easy to just pick up and read. It is comprehensive. Interesting exercises in Chapter 5. Lots of good references.

[VM] *The Visual Mind*, Michele Emmer, ed., MIT Press, 1993.
A collection 36 papers dealing with mathematical aspects of art. Several papers are relevant to this book. The paper, "Interlace patterns in Islamic and Moorish art" includes Cayley diagrams of wallpaper groups.

[VS] *Visions of Symmetry*, by Doris Schattschneider, W.H. Freeman and Co., 1990.
The comprehensive source of Escher plane tilings. Pages and pages of fascinating pictures. Description of Escher's own classification of patterns.

Index